Cut and Run
Loggin' Off The Big Woods

Mike Monte

4880 Lower Valley Road, Atglen, PA 19310 USA

Dedication

To my father, Richard "Boyd" Monte

Acknowledgements

I thank the people who have shared their old photos and stories with me through the years, and the photographers who used primitive equipment to chronicle early logging days in the Lake States.

Library of Congress Cataloging-in-Publication Data

Monte, Mike.
    Cut and run: Loggin' off the big woods/by Mike Monte.
      p. cm.
    ISBN 0-7643-1529-3 (pbk.)
    1. Logging--Michigan--History. 2. Logging--Minnesota--History. 3. Logging--Wisconsin--History. 4. Logging--Environmental aspects--Michigan--History. 5. Logging--Environmental aspects--Minnesota--History. 6. Logging--Environmental aspects--Wisconsin--History. I. Title.
    SD538.2.M5 M66 2002
    634.9'8'0977-dc21
                      2001008280

Copyright © 2002 by Mike Monte

All rights reserved. No part of this work may be reproduced or used in any form or by any means— graphic, electronic, or mechanical, including photocopying or information storage and retrieval systems— without written permission from the copyright holder.

"Schiffer," "Schiffer Publishing Ltd. & Design," and the "Design of pen and inkwell" are registered trademarks of Schiffer Publishing Ltd.

Designed by "Sue"
Type set in Americana XBd BT/Korinna BT

ISBN: 0-7643-1529-3
Printed in China

Published by Schiffer Publishing Ltd.
4880 Lower Valley Road
Atglen, PA 19310
Phone: (610) 593-1777; Fax: (610) 593-2002
E-mail: Schifferbk@aol.com
Please visit our web site catalog at
www.schifferbooks.com
We are always looking for people to write books on new and related subjects. If you have an idea for a book, please contact us at the above address.

This book may be purchased from the publisher.
Include $3.95 for shipping. Please try your bookstore first.
You may write for a free catalog.

In Europe, Schiffer books are distributed by
Bushwood Books
6 Marksbury Ave. Kew Gardens
Surrey TW9 4JF England
Phone: 44 (0)20 8392-8585; Fax: 44 (0)20 8392-9876
E-mail: Bushwd@aol.com
Free postage in the UK. Europe: air mail at cost.
Please try your bookstore first.

Foreword

For more than 15 years, as I've edited his various articles for *The Northern Logger*, I have enjoyed (and to be honest, envied) Mike Monte's sophisticated wit and writing skill. Combined, they make him both an entertaining and provocative chronicler of human behavior and events.

Where he acquired this rare ability is a mystery to me, since he has none (and I mean none) of the big city trappings that some of us tend to associate with insightful social commentary and observation.

As much as I've enjoyed Monte's work for *The Northern Logger*, I see that he's saved some of his best writing and research for this book. Having read more than my share of logging history over the years, [this book] stands out in the way the author combines personal knowledge, experience and thought with exhaustive research, weaving it all into a fascinating narrative that is hard to put down.

Eric Johnson, Editor
The Northern Logger

Contents

Introduction 4

Chapter 1. Lumberjacks and River Pigs 6

Chapter 2. Logging and the Railroads 41

Chapter 3. Padus, Profile of a Sawdust Town 114

Introduction

It was an unprecedented rape of Mother Nature, an almost complete change of the landscape across the northern ends of three states, and the social and ecological changes will be felt for evermore. It was a time for empire building, and some of the lumber companies born back then still exist today. Others had a brief and glorious period of profit and faded from sight, leaving only the rock and concrete foundations now overgrown in forgotten corners of the landscape. Towns grew like weeds around sawmill sites, often built by the lumber companies. These towns, sawdust towns, like the mills that nurtured them, often died when the last tree was cut. The people living there called it "cut and run."

The big lumber companies were the major landholders in most of the Upper Great Lakes. As forests were turned into lumber, lumber companies whose sole mission was to turn cutover lands into homesteads created land companies. Easy payment plans put families onto their future farms, and years of agonizing work removing rocks and stumps yielded some scant acreage whereby a family was raised.

When the big push was over, the lands, now known as the "cutover," became a national scandal. Homesteaders were stranded both physically and economically when lumber companies moved on. Logging camps and the armies of hungry men inhabiting them had been the only market for their produce. Working in the logging camps kept the homesteaders in cash through the winter months. When the forest was gone, in almost all instances, the homesteader was out of business, and after abandoning those years of backbreaking work clearing the stumps, the forest began reclaiming the land.

In spite of all this, it is a period in Lake States history that is remembered with fondness, a glorious time of making fortunes or spending your youth in a devil-may-care orgy of hard work, strong drink and hopefully some women with less than shining morals. Even if most of the communities created out of the forest disappeared, some didn't, and the infrastructure created during the big cut still serves the citizenry today. Many residents of the Upper Great Lakes can trace their ancestry to a grandfather or great grandfather who was drawn to the pineries or great hardwood stands of the Northwoods, and stayed on to raise a family. The biggest contribution to the demise of the original northern forests can probably be found on the prairies and in many of our great cities.

We were a growing nation in the midst of the industrial revolution when the forests of the Upper Great Lakes were cut. Our cities were expanding in huge leaps and the Great Plains, the treeless prairie, was filling with homesteaders. In both instances, people needed houses, and doubtless, thousands of families still live in homes constructed from the pines and hardwoods cut in Minnesota, Wisconsin and Michigan. If the landscape was scarred, so be it. (Try telling a housewife, who had spent the last five years raising kids in a sod house, that she couldn't have a new frame house because the pine was being over cut!) The needs of the nation were met, and the mindset was twofold; the rape of the woods was necessary for the well being of the nation and the cutover lands would turn into productive farms capable of helping feed the increasing population. It is easy to look back and see the folly in destroying a resource, but then again, hindsight is always 20-20.

This work is not a precise history of the logging and sawmill industry in the Upper Great Lakes. That would take many volumes, if indeed it would be possible today considering the short life of many of the logging companies and the lack of any accurate records. It is a brief look at what happened to the forests and people of yesterday. Much of the material within is from the northeast corner of Wisconsin, and while it is not a comprehensive look at the subject, it is a representative look at the life and times of the period. The people in the old photos may be different, but the same stories and images exist throughout the upper Great Lakes region.

Pine was king in the Lake States, and pines were the first trees logged when the loggers poured in from the eastern states. Loggers from New England had already created vast fields of stumps in New York, Maine, Massachusetts and Pennsylvania. The big mills and mill owners moved west. There were lumber markets to fill, and those markets were for pine lumber.

Why was pine so desirable? Pine is easy to saw, dry and plane into a smooth board. It is easy to nail, light and easy to handle. It was the first choice of homebuilders, and what was most important, it worked with the existing transportation system.

Rivers provided the route to mill and market for pine logs. Pine logs float high and easy in the water. Denser hardwood logs do not. The river systems of the Lake States were a natural highway for pine logs. To make things even handier, stands of pines often proliferated along the river bottoms, making the sleigh trip to the river shorter.

Sawmills sprouted downstream from the pine stands along the larger rivers. Cities like Oshkosh, Eau Claire, Peshtigo and Stevens Point became sawmill centers, waiting each spring for the log drives that supplied the sawmills that were the economic engine of the new communities.

After the logs, millions of logs, were milled into lumber, they were often made into rafts, and the rivers once again served as transportation for the pine. Men rode and steered the rafts down rivers like the Wisconsin to markets like St. Louis., where the rafts were broken up into boards and sold to the consumer. The Great Lakes were another transportation system that enabled the exploitation of the pine. Steamers could pick up lumber at numerous ports around the lakes and deliver it to transportation centers like Chicago, where railroads could move the pine lumber to markets in the east or west. The demand for the lumber was there, and the loggers and sawmill men were willing and able to meet that demand. And, there was always another stand of pine waiting to be cut. It was an inexhaustible resource, just ask anyone.

The men who built the sawmills and turned those many millions of board feet of pine into lumber came to be known as "lumber barons." And certainly, many of them fit the description. After all, just like the royalty of Europe, they controlled thousands of acres of land and the people living on that land. Some were loved, some were hated. Probably, many of these men didn't care if anyone liked them or not. They were there to make money as quickly and efficiently as possible. Most of them were quite good at what they did. Some turned their logging fortune into other businesses that made more profits and others lived an extravagant lifestyle, spending their money just like the men they employed in their camps, albeit in a more luxurious fashion.

The winter cut is over. It's time to celebrate and enjoy being young and strong and a veteran of a winter lumberjacking in the big woods. A few drinks and a photograph to show future grandchildren seems appropriate.

Chapter 1
Lumberjacks and River Pigs

Like the rest of the population, the men who worked in the big woods were not homogenous. There were men who lived solely for the day, hard drinkers and brawlers who labored from "can see to can't see" through the cold winter months only to blow the winter earnings on a rip roaring drunk come spring breakup. Others spent a season or two for the adventure, moved on to other pursuits and were able to tell their grandchildren of the days when they worked in the big timber and rode logs down the river. Farmers from the southern counties of Wisconsin, Minnesota and Michigan, having a family to take care of the farm work, spent the winters in the Northwoods, renting out themselves and their team of horses to skid logs, adding some cash to the family larder.

The photographer's studio drew young lumberjacks in to record the moment and give the folks back home a look at what a dashing figure they had become. The gentleman on the right is wearing stagged pants. Stagged pants were worn because long trousers caught on brush and became wet and heavy in snow or water. Less pant leg meant lighter pants. The pants were shortened with an axe. The pant leg was twisted and held on a stump. The axe made a neat cut that wasn't jagged. The inseams might not be close to identical, but then who the hell cared?

This photo of a logging camp in northern Oconto County, Wisconsin, was probably taken in the 1880's or early 90's. It wasn't a typical day in the camp. The photographer was there, or there wouldn't be any sleigh loads of pine logs sitting around, or that many idle men either. The logs would have been sliding to a landing on the Oconto River and the men would have been out in the woods piling up more logs for the sleighs to haul. It is a great shot of what an early logging camp looked like. It appears that the bunkhouse and cook shack are on the left and connected by a roof with an open area between. The barn is a work of art, constructed of logs and sturdy enough to last for decades, even though it was probably abandoned when the timber was cut off. The loggers are using oxen and horses. In later years, oxen were considered too slow and horses were more common. A cutter with a single horse is to the right, and was likely used by the woods boss to check on the men.

Logging was a labor-intensive industry in its early days. Unlike today when mechanized logging crews consist of a few men and some high-tech machines, each task of the logging process had men assigned who did that job only day in and day out. They might "graduate" to a new job requiring different skills, but they usually stuck to the assignment given them. There was good reason. Logging had to proceed on a precise schedule that was determined by the weather. Winter was the season. Snow and cold enabled loggers to gain access to low ground, build ice roads to allow the use of logging sleighs, and in the case of river drives, the deep snow provided the spring floods that floated the logs to market.

A typical logging operation started, of course, with the purchase of the standing timber within sleigh haul distance of a sizable river. In many instances the purchase was forgotten and finding a good stand was enough. Ownership of a timber stand only became an issue if the logging operator was caught logging it. A logging camp was built, which consisted of a dinner shack, bunk house, and office, which usually doubled as a small store to provide necessities like leather mitts, socks and chewing tobacco. A blacksmith shop was usually imperative for keeping equipment in working order, and barns were needed to house horses and/or oxen.

Chewing tobacco is still used by loggers and plenty of other people today, but in the old logging days, everyone chewed. Brands like Standard and Peerless were popular, but the favorite seems to have been Spearhead. When looking at old records from the Page & Landeck company store, almost every charge had a pound of Spearhead listed. According to a gentleman who bought an old store in a logging town and found the records, chewing tobacco was ordered by the ton!

Typical logging crew and typical logging camp is an apt description. The men are of varying ages and backgrounds, and they will spend their few non-working hours in the tarpaper-covered bunkhouse behind them. Look at the roof of the camp and you can see where the warmest part of the building is. The skylight on the roof was a necessity of camp life. When the room warmed up too much in the evening, the skylight was opened and the room cooled off. It was also handy to let out the smells of wet wool, tobacco juice and the second-hand odor of the beans enjoyed at supper.

The photographer needed some loggers for a picture postcard, and these guys were picked. Notice the matching outfits. Were they all good friends who like to dress alike? Not likely! But, it is likely that the camp store got a deal on wool jackets and sold the jackets to these fellas. It was a painless way to shop. The cost of the jackets could be taken out of their pay come spring.

Maybe some photographers had more sense of humor than others. These lumberjacks are clowning for the camera, and not standing neatly in a line, like many of the early logging photos.

A crew was gathered up and the cutting and skidding of logs could begin in the fall before freeze-up. Sleigh roads were started by "swampers" who worked with axe and grub hoe. Logs were decked with a team of horses and a "jammer," usually an a-frame jammer with a double line, but sometimes with a more dangerous and less efficient single line. Men worked with a crosscut saw and axes, falling timber and bucking or cutting the logs to lengths. Sometimes, there were three men working together, with two men sawing and the third limbing with an axe. These were considered top men in the logging camp pecking order. It took skill and muscle to be a good sawyer. Skill in being able to fall timber efficiently and put trees where you wanted them and skill in staying alive. There are plenty of unskilled woodsmen in unmarked graves who didn't pay attention to falling limbs that plummet from the sky, heavy end down like a dart, and with enough force to cave in the thickest skull. Men who pulled crosscut saws six days a week built up huge muscles in their shoulders.

These three fellows were cutting this timber around 1900. The older man with the mustache is carrying an axe and a measuring pole. When the younger guys fall the tree, he walks down the tree and measures the logs. The eight-foot pole is marked in two-foot increments. The logs will be cut either 8, 10, 12, 14 or 16 feet long, depending on the crook of the tree, the taper of the trunk and the amount of limbs. He will also limb the tree ahead of the two sawyers working the "misery whip." If you look to the right in the photo with the standing, notched tree, you will see the kerosene can. Kerosene was used to wash pitch off the saw, making it easier to pull. Often, one of the sawyers would carry a half-pint whiskey bottle of kerosene in his back pocket, stoppered with a rag. This was used to lubricate and clean the saw while making the cut.

Oxen were used in the early days of logging, and were gradually replaced with horses, although die-hards still used teams of oxen. There was many an argument over the use of horses or oxen. Both animals had their good points. Horses were faster than oxen but required more care. Horses will eat oats until they almost explode, but oxen, and mules for that matter, will eat what they need. Horses required constant care. If a horse wasn't being worked, he had to be put to pasture. Oxen, on the other hand, were often left in the woods after spring breakup and would fend for themselves. Wearing a collar with a bell, the teamsters would look for them in the fall, usually finding them by listening for the bell.

This teamster is holding his goad stick, the method used to drive oxen. Good men with oxen were easier on their animals than others. Teamsters who marked and bruised their beasts of burden were looked down on by most of their contemporaries. A skilled man got the work done and was gentle to boot.

> *My father, who worked as a sawyer in logging camps at the tail-end of the era in the late thirties and early forties, said he weighed 135 pounds soaking wet. He claimed his shoulder muscles were so developed he couldn't wash the back of his neck unless he rolled the washcloth and grabbing each end flipped it over his head, taking off the grime with a sawing motion.*

Swampers were a notch or two lower in importance than sawyers. Working with an axe or brush hook, they cleared skidding trails for the teams of oxen or horses that skidded logs to the decking crews on the sleigh trails. Decking crews piled the logs parallel to the trail so the loading crews could put loads of logs on sleighs once the ice road was built to the decking areas next to or on a frozen river or lake. The logging operation was supposed to work together and move through a timber stand in unison. It took time to set up a jammer and nobody wanted to backtrack, especially the woods boss who coordinated this effort. As the trees were cut into logs, the skidding crews skidded them to the trail and the decking crews put them into piles, each crew keeping the other busy and each crew not falling behind. Like a worm eating through an apple, the loggers ate through the forest.

The camp mascot is posing on a nail keg, but there are some other interesting things in this photo. The small sleigh on the left brought hot food to the jacks out at the job site. The crosscut saw on the right is held in a vise or clamp used for holding the saw steady while it is filed or sharpened. Most camps had a filer who knew how to make a crosscut cut frozen timber, saving lots of muscle power and upping production. The young man holding the double-bit axe probably has one side razor sharp and the other side dull. The dull side was used for limbing in extremely cold weather, to prevent taking a chip out of the axe. Ask any good axe man and he will tell you that all axes aren't equal. Some were junk, and some were high quality and owned with a certain amount of pride. Plumb axes were considered good quality, and Kelly Hand-Made axes were the Rolls Royce of axes. An old Kelly is still a prize possession with some loggers who know axes. Mine is leaning against the wall behind me as I write this, and was purchased at a garage sale from grandchildren who didn't know that Grandpa's axe was special.

This crew was probably clearing landings or making an ice road. All but one. The gent on the left with the Winchester is probably management. He is wearing rubbers with wool socks, but how many loggers would be caught dead wearing garters?

A team of horses could easily handle a pine log of this size. This teamster is using tongs to hook up to the log. They were quicker than hooking a chain. The hook up was always done at an angle, so the log would roll a bit when the horses started, making it easier for the team. If the log hung up against a root or stump, the tongs could be rehooked with a different roll to clear the obstacle.

Logs were loaded from the decks in the woods onto sleighs that transported the logs to the river, and another log deck on the bank or sometimes on the ice covering the river, to await the spring thaw. A good teamster could turn the air blue with cussing for at least a half-hour, and not overwork any four-letter word. It was part of the trade. Horses who had been trained by a colorful talker didn't know a command unless it was followed by a "goddamit" or a "you sonavabitch." Skilled teamsters could move logs without hurting or overworking their animals. Getting a good run on a steep hill or keeping the team ahead of a sleigh coming down a steep grade were all part of the job. Perched on a bag of oats on top of a load of logs held together by chains was testament to the teamster's faith in the top loader.

While cussing might be a taboo in polite circles, it was a healthy release of frustration in the logging camps. Men who were very good at it became legendary. I have heard that a good cusser can turn the skill into an art ofrm. This may be a slight exaggeration, but I have heard about and heard myself some fantastic profanity.

Old timers around my area told about Earl Armstrong, an early teamster. A fellow worker said that Earl was training a horse to harness, and the animal was proving to be either stupid or strongheaded. According to the eye-witness, Earl fell to his knees, raising his hands in the air like a tent-show preacher, he bellered at the horse, "May the Good Lord come down in the form of a spike-toothed drag and land in the middle of your back, you no-good sonovabitch!"

Some of these skills were passed down to the next generation of loggers. About thirty years ago, I was logging a tract of timber with my friend Doug Champine. Champine, who has now passed on, was born at least fifty years too late. He would have fit into the early logging days quite easily. The logging was being done on land that belonged to a turn-of-the-century resort. A caretaker lived there and watched over the property during the winter months when the owners were back in Chicago. We were cutting big, mature maple, and everything would have been perfect, except the skidder operator didn't always get to work every day. Logs littered the ground for acres, and there was a danger they would be snowed over and not found until summer, spoiling the logs. Champine unloaded for at least ten minutes on the hapless young man, and was at the pinnacle of the art of cussing. Myself and the caretaker were spectators to the event. I was proud of Champine's performance and would have rolled in the snow laughing if it wasn't for the fact that the recipient of this harangue was also a friend. I looked at the caretaker, whose name was Wally, and saw his jaw was hanging agape and his eyes were opened wide. When Champine had finished and stalked off into the woods, Wally looked at me and said, "You know, everybody cusses. I do. You do. Just about everybody does, but that man is going to hell for it!"

This is a good shot of a loading operation using an a-frame jammer. The whole logging crew has come out of the woods to be in the photo, but there were probably four men loading the sleigh; two hooking pups, one driving the horse that raised the log, and the fourth top-loading.

HAULING THE BIG LOAD AT MONTE'S CAMP CRANDON-WIS

It could get a little dull around a logging camp. Maybe, that is why woods bosses decided to build a show-off load. It took some prime logs, usually selected from the woods-run cut, then a photographer was hired to record the event for posterity. Hooking up a team of horses that was dwarfed by the load they were supposed to haul was also mandatory. The result was a photo like this. The team probably did haul the load to the landing or the mill, but it was sure to be a downhill run. This load was put together by my Great Uncle George, a woods boss with the Page & Landeck Lumber Co., Crandon, Wisconsin.

The top loader was the undisputed boss in the loading operation. A jammer was setup along the sleigh road on the opposite side of the log deck. A cable with a yoke, each end fitted with a pup-hook, was strung through the top of the jammer. Two men hooked pups, another drove a team of horses that pulled the logs from the deck, usually up a pair of skids or logs leaning against the sleigh. The top loader supervised the placement of the logs and the chains that bound the load together. Teams used in loading were trained differently than those used in skidding or pulling sleighs. A loading team had to pull the log over and up on the skids, but they also had to lift the log and hold it in the air until it could be swung into place by the top loader or other workers on the crew. Logs heavier than the team, of course, couldn't be held for long, and that was part of the skills used by the top loader. Heavier logs were selected for the bottom of the load and lighter logs for finishing off the load. Loads had to match the terrain; hilly and rough, lighter loads; smooth and slightly downhill to the landing, top her off high!

Decking logs and loading sleighs, like most jobs in the woods was dangerous. A hook that let go dropped hundreds of pounds out of the air. A man working a loading or decking operation not only had to know where his feet were at all times, but he also better know what was over his head. Standing under a log swinging on two small hooks was not a bright idea. Smart workers gave the job a lot of respect and always made sure they didn't expose themselves to danger any longer than necessary. Still, there were plenty of broken legs, smashed hands and more than enough deaths.

Not to be forgotten in the complicated system of logging back in the old days was the road monkey. These pictures of sleighs heaped with heavy logs don't usually show the reason they were able to be pulled with two, or maybe four horses. Logging sleighs ran on ice roads that were built by the road monkeys. While they didn't have a title that garnered much respect, they were an essential part of the operation.

At the beginning of each logging season, as soon as the freeze set in, sleighs equipped with water tanks were used to ice the roads. The tank was usually a wooden box, and the tank was fastened to a sleigh. Holes in the back of the tank spread the water over the roadbed, and eventually an ice road was built up, level, smooth and slippery. The next step was cutting grooves for the sleigh runners. Special cutters adjusted to the width of the sleigh runners were pulled behind a team of horses. All the sleighs used by a logging camp had to have the runners the same gauge or distance apart. The camp blacksmiths usually made all or most of the equipment used by a logging company, so it wasn't a problem. The lowly road monkeys had to constantly work on the sleigh roads. While the loggers were snoring their weariness away, they were out in the cold night, dipping water with buckets into the water tank and spreading a new layer of ice. Some road monkeys did work during the daylight hours, however. The lowest man on the totem pole would walk the road and remove any bark, horse apples (manure) or any other material from the ruts. Other men might be employed sanding the hills so that the road wasn't slippery. When sleighs were pulled downhill, the loads of logs could sometimes slide faster than the horses could run, overrunning the team, which usually meant dead horses, spilled logs and sometimes a dead teamster. Most teamsters were ready to jump to safety from a runaway sleigh, but nobody wanted to jump too soon if there was a chance to save the team and keep the logs on the sleigh. Some waited too long. Another hazard was sleighing logs across a frozen lake. A spring could make a thin place in the ice, and more than one teamster made a leap for solid ice as his loaded sleigh and team broke through, headed for deep water. Almost all lakes in the Lake States have a legend about a load of pine logs that is somewhere on the bottom, along with a team of horse skeletons still in the harness.

The late Oliver Campbell told me about the longest night of his life, and that included the times spent in the trenches of World War I. Oliver was a teenager working on a decking crew near Hiles, Wisconsin around 1910. A crewmember who didn't pay attention to the stability of the logs in the deck had a big log roll over him, literally squeezing the life out of him. The body was wrapped in a canvas and hauled into Hiles. The family of the deceased was notified, probably by telegraph. They were coming to pick up the body with a sleigh, as the few roads that existed were never plowed in those years. The custom, back then, was that a dead body was never left alone until it had been properly buried. Oliver, who said he was the youngest member of the crew and likely the least necessary, was given the job of sitting with the corpse in the Hiles Town Hall. Imagine being a teenager and spending the night alone in a room illuminated by a kerosene lantern, with a mangled corpse that only hours before had been a talking, laughing member of your crew! A long night indeed. The old town hall is now a sawmill office, and I have never looked in the back room without thinking of young Oliver and that canvas-wrapped corpse.

All of these stories aren't legends. One spring, right after ice-out, I was canoeing on Lake Metonga, near Crandon, Wisconsin. The lake was smooth as glass, and the water was clear down to at least twenty feet. There on the bottom was a pile of logs protruding from the sandy bottom, piled neatly, and obviously not put there on purpose. Was the team lost? Did the driver die? Those answers are lost to time, but needless to say, the logs never made it to the head rig at an Oshkosh sawmill.

Sometimes called a water wagon, even though it was on a sleigh, the work accomplished with these crude water tanks kept the loads of logs sliding to the landing. Usually, the men who kept the sleigh roads iced and grooved for the runners, worked after dark, when the rest of the men were in the bunkhouse. It wasn't an easy job. Water was dipped from a hole in the ice of a river or lake, and trip after trip was made to keep the road in shape. An old logger, who had done this kind of work, told me that they returned to the dinner shack after the other men had left for work. After breakfast, they went to bed. He also said that there was a bonus. When dipping water, they would usually dip some brook trout, which they set aside and had the cook fry up for breakfast.

These would have been normal loads of pine logs. There is about 2000 board feet of white pine on these sleighs. Loads like this could be pulled over hills or wet swamps, and they were easy on the horses.

Old loggers always said that cold weather didn't stop them from logging. A close look at this photo gives testament to a cold day in the woods. The horse's whiskers are frosted and so are the teamster's. It was cold enough to wear that heavy buffalo hide coat too. These logs were cut near Tipler, Wisconsin, and were probably floated down the Pine to the Menominee and down to one of the mills at Menominee, Michigan or Marinette, Wisconsin.

At winter's end, a successful logging operation had reached their goal in board feet and had the logs decked on the stream or a lake that fed a river. Sometimes, the logs were decked on the ice, sometimes on the bank or both. Now it was time to wait for the thaw. As the winter snows melted, filling the rivers behind the dams, the driving crews were selected. Not everybody was needed for the spring log drive, and not everybody wanted to go on the drive either. For the farmer who had spent the winter away from his family, it was time to say goodbye to the logging camp and prepare for spring farm work, not to mention spending the nights with a warm woman who smelled a lot better than his bunkmate of the last few months. Undoubtedly, some of the adventurous youth who wanted to experience the life of a lumberjack already had enough adventure to last the rest of their lives and opted out. For others, it was the crowning moment of the logging season.

This huge pile of pine logs on the Wolf River in Langlade County, Wisconsin was called the "Big Rollway." For a number of years, until the pine played out, logs were sleigh hauled to the rollway and unloaded over the edge of the riverbank. According to the caption on the photo, there was 1,500,000 board feet of pine piled 182 feet high. The pile is so huge that big logs look like matchsticks. Imagine the danger to the men armed with peaveys and pike poles that had to roll these logs into the river when the spring drives started.

This old postcard was called, "A Log Landing On the Oconto River." The time would be in the 1880's or 90's, and obviously it is "visitor's day." There is one working jack in the photo, over on the left side holding a peavey. The rest are visitors, probably affiliated with the lumber company, and on a lark upstream to see where all the wealth came from. They came well armed with a variety of rifles.

The Big Jim Rollway on the west branch of the Wolf was piled with 1.5 million board feet of pine. According to the inscription on the photo, there were 140 men on hand to roll the logs into the river. This is the spring of 1908.

River pigs on the Wolf River wait for the dams to open and give them a log-floating flood. The rear crew did most of the heavy work. When the water level dropped, logs were stranded on the bank and on sand bars, and the rear crew rolled and grunted them back into the main stream, where they would continue their journey to the mills at Oshkosh. Take a close look at the faces, and you will see a number of Native Americans. The logging industry had very little prejudice with its hiring practices. If you could do the work, you were paid the same as everybody else. The old river man Whitehouse, in his autobiography, said the Menominees and Chippewas were great, nimble river pigs, and their labors on the log drives were appreciated.

What is it that being in near-frozen water all day, sleeping wet under a canvas in early spring temperatures and eating beans three times a day that brings back happy memories in some and revulsion in others? To each his own! Fortunately for the mill owners downstream, there were plenty of men who thought a log drive was a great adventure and not a hardship. Without them, the mills in Oshkosh, Eau Claire and other sawdust cities would have remained silent, and the great fortunes of the timber barons would not have existed.

The posing was over in this shot, and it was back to work. The bateaux is carrying men downstream to float some more stranded logs.

This shot of a log drive on the Wolf River gives a good idea of the volume of logs moved in a drive. Remember, the majority of the logs in the drive had already passed this spot. The haphazard position of the logs indicates that they were left on the banks after the flood passed. When water was released from the dams upstream, the rear crew would roll the stranded logs off the banks into the flooded river. Tents were used to feed and bed down the river men.

Many of the big rivers like the Wisconsin and Chippewa didn't require as many dams as the smaller rivers. They were a big body of water without dams. Dams on the tributaries were used to create floods that kept the logs floating high on the big river, and as the pine was cut on the larger rivers, hungry lumbermen looked at smaller rivers and their pine crowded banks. The upper Wolf River, in northeast Wisconsin, was considered too small by many of the lumbermen of the time to handle a log drive. The lower Wolf was easy to drive on, no rapids and a deep easy current. Logs were boomed and sorted on Lake Poygan according to ownership and pulled in rafts by tugboats to the various sawmills in Oshkosh. Ownership was determined by a stamp in the end of the log with the company trademark or brand. This tool was actually a hammer with the symbol extruded on the business end of the hammerhead. As the easy trees became scarce, the fear of mills without logs caused eyes to turn north to the upper reaches of the Wolf River drainage.

Philetus Sawyer was an Oshkosh mill owner and a U.S. Congressman, and later on, a U.S. Senator. Sawyer did an onsite inspection of the upper Wolf in 1869. He judged that the Wolf and its tributaries could be used if the river was "improved." He acquired a charter from the state and federal governments that gave him full control of the Wolf and its tributaries. The term "conflict of interest" wasn't in the government vocabulary in those days. He formed the Keshena Improvement Company and hired men who knew how to make dams and use dynamite. As the Keshena Improvement Company worked its way upstream, millions of board feet became available for the mills in Oshkosh. Sawyer also acquired land that he could resell. Logging companies bought the standing timber and paid the Keshena Improvement Company for the use of the river. When the logs arrived in Oshkosh, Sawyer was able to buy all he needed to keep his saws in wood. He made a lot of money and forever changed the Wolf River.

> On October 8, 1871, the Peshtigo Fire destroyed the village of Peshtigo, on Lake Michigan, at the mouth of the Peshtigo River. Over 1000 people lost their lives. The Peshtigo Fire didn't get a great deal of press, because the Chicago Fire occurred on the same day. Newsmen, like today, covered the story with wider reader appeal. But, the Chicago Fire provided a huge market for lumber, and the story is that much of the timber floated down the Wolf was used to rebuild Chicago.

This is a famous spot on the Wolf River called Gilmore's Mistake. When surveying the Wolf in 1869, Mr. Gilmore went ahead of the party and found the river gurgling out of the ground. He hurried back to the main party and said they had reached the end of the river. The amazed crew went to look. What they found was the result of a tornado that blew trees across the river. Nature can play a few tricks. The thick mat of tree trunks caught needles and other debris and soil formation began. Trees seeded on the remains of their dead ancestors, and the Wolf ran under a natural bridge resembling the ground. The hapless Gilmore was forever immortalized. This is also another view of the backbreaking work the rear crew had to perform to get all of these logs back into the river.

Controlling a river was a full-time job. Men worked year-round to keep the dams in shape and to build new dams as they were needed. It was an acquired skill. Using the materials available, a good dam crew could use brush and dirt to build out from the stream bank toward the middle of the river where the gates were constructed that controlled the water. Raise the gate and water would pour through. If there were logs behind the dam, men with pike poles could push and pull the logs so the ends passed through the gate. When more water was needed, the gates were lowered until a head of water built behind the dam. If the drive had already passed through the dam, or if a dam was on a tributary, the water was still needed downstream to float the logs. Dam tenders would open the gates at specified times, creating an artificial flood that floated logs off of sandbars and snags downstream. The men who poked the logs through the dams and tried to stay with the front of the drive were called the "jam crew." The "rear crew" worked the back of the drive, and toiled hard to roll logs off sandbars and riverbanks when the flood had passed. Both crews were usually referred to as "river pigs." Sometimes they missed a few logs, however, and if the log spent a season in the river it became waterlogged, sank and was then referred to as a "deadhead." Logs left in water last forever, and there are still deadheads in the streams of the Great Lakes region.

A bateaux runs down the Wolf. The bateaux were used on river drives from Canada and through the eastern states. It is a contribution of the French Canadian loggers, and they were a stable vessel. Bateaux carried men, tools, food, and everything else needed by the men on a log drive.

A mill yard in the Oconto River drainage with the wannagan or floating cook shanty tied up, as well as two bateaux.

Running whitewater in a bateaux could be quite sporting and dangerous, but the vessels were too heavy to carry for any distance, and what the hell, life was supposed to be dangerous on the river. A. J. Kingsbury, a photographer from Antigo, Wisconsin took the original shots of the log drives on the Wolf River, and this photo of a bateaux running the Dells of the Wolf is amazing photography when you consider the type of equipment he was using. Certainly, the run and the shot were planned.

The men lined up for a photo op on the logjam. The river men might not have cared greatly if there was a jam, but to the owner of the logs, like today, time was money. The Dells of the Wolf were notorious for hanging up logs and stopping the drive. Breaking the jam was a dangerous job, and they are still looking for the bodies of some men caught in front of jams that let go unexpectedly. Dynamite was often used to blow the key log, letting go a roar of water and pine timber. Sometimes brave or foolhardy men had a rope tied around their waist and walked out on the jam, rolling logs clear. The safety line was also a good way to recover the corpse. Dead loggers were usually buried on the bank with their boots overhead as a marker. The Dells of the Wolf were such a nuisance that the Keshena Improvement Company dynamited the rock back 20 feet or so on each side of the river.

As the log drives headed downstream, usually, the rivers widened and became tamer. The cooks could get their gear out of the bateaux and set up their kitchen on a raft that could stay ahead of the drive. Meals were cooked on the raft over a stove, making life a lot easier. This cook shanty on a raft worked the Wolf River in the 1890's. There was a building for sleeping and the other for cooking.

This raft was used on the Oconto River sometime before 1900. It is obviously the end of the drive, and the fairer sex with their children are making a visit.

This floating cook shack was probably on the Oconto River, or possibly the Menominee. It looks like the evening spuds are being peeled.

The Gardner Dam on the Wolf River. A dam like this was a major building project for the Keshena Improvement Company, and made the drives on the Wolf possible. The Gardner Dam let go in the 1873 season during a torrential rain. Eight men rushed out to open the gates so the earth and brush bulkheads wouldn't wash out. They had just started to open the gates when the whole bulkhead let go putting the men in the river. Four men survived, and the other four drowned.

Men who worked with rivers became very adept at changing flows and raising water levels using the material at hand. A Wisconsin lumberman changed the course of history during the Civil War in the Red River Campaign, in Louisiana, using the skills he had learned floating Wisconsin pine to market.

Admiral David Porter, along with ten of his gunboats was stranded by low water above the falls and rapids at Alexandria. Porter was faced with possibility of burning or blowing up his boats to prevent them from falling into enemy hands. Lieutenant Colonel Joseph Bailey, who had put a hold on his logging career to fight for the Union Army, dissuaded him from destroying his small fleet. Joe Bailey knew that the Red River could be made to flood and float Porter's boats. The Red River was only three and a half feet deep and had to be seven feet deep for the boats to float over the rapids to the deeper water below.

Bailey had the help of over 1,000 infantrymen from the Union Army, and some of them from Maine, who knew how to make a logging dam. They constructed wing dams from either bank and plugged the middle with sunken barges, a distance of almost 800 feet! It took only a week to place the trees, brush and rocks that formed the dam. When the water was over seven feet deep, the sunken barges washed out, creating a torrent that swooshed four of the boats through to safety. Another dam constructed upriver floated the other six boats to safety. This dam, now built with educated dam builders, was constructed in only three days!

It is hard to believe that millions of feet of pine timber floated, bumping and grinding over these rocks! This photo of the Langlade Dam on the Wolf River gives a good idea of how a dam worked. When the logs were floating high behind the closed gates of the dam, the gates were opened. Men worked frantically with pike poles and peaveys to turn the logs so they would float through the gates and ride the flood as far as possible. Dams built in stages could move the logs in stages downstream to deeper water.

The Lake states are filled with place names that are followed with the word "dam." While there may not be a dam there anymore, the name persists. In other instances, dams were retained, rebuilt, and still hold back a lake that is now a recreation area. The men who built and used those dams changed the characters of the rivers they controlled, with untold yards of soil from the banks washing downstream every time a flood was created, many of these streams look very little like the stream before the dam. In some places, streams have cut their channel deeper, big trees grow on the banks and it would be hard to tell that any massive disruption had ever occurred. It would be unthinkable to approach state government with a plan to dam and float timber down a stream today!

Every logging camp had one very important employee. His skills could guarantee a happy and productive work force and a low turnover in the highly skilled portion of the men. He often commanded more respect than the woods boss, the scaler, and no doubt the owner of the operation. He was the cook!

This cook in a camp on the Oconto River, near Lakewood, Wisconsin, had two young bull cooks or cookees, sometimes called the chore boy. The cookee on the far right is holding the "Gabriel" that was used to bring the men out of the woods to eat. The men on the far left, judging by their dress, are management, up from the headquarters at the mouth of the river. The cookees had to be up before anybody in camp to get fires started and to wake up the men. The teamsters were usually rolled out of bed first, because they had to feed their animals before breakfast. This usually meant they ate breakfast before the rest of the crew. Cookees could be very unpopular in the camp if they made too much noise waking the teamsters and woke up the rest of the bunkhouse prematurely.

The dinner shack with the table set must have been a welcome sight to men who had worked hard all day in the cold and snow. The cook slept in the dinner shack, and if you look at the back of the room, you will notice a mirror and a razor strap, obviously the spot where the cook washed and cleaned up in the morning. The granite wear coffee pots are on the shelf waiting for the gallons of hot, black coffee that would be consumed at the next meal, and everything is tidy.

When a man worked for low wages, slept in a straw bed, and was often thought less important than a good team of horses, his daily meals became the one thing in his life that he looked forward to. A logger burned an enormous amount of calories working hard in cold weather, calories that had to be replaced three times a day. A good cook, and his assistants or bull cooks (also called chore boys or cookees), had better know how to give a lumberjack a breakfast that would last until lunch time, which was usually served hot on the jobsite, and he better know how to give him a supper that put him in bed with a contented feeling. Many are the jacks that packed their bag or "turkey," threw their axe over their shoulder and started hiking down the skid road, looking for a camp that knew how to feed their men.

The late Junice Peek, a storyteller from Lakewood, Wisconsin, told of a camp cook who wasn't very talented. The men were grumbling and the cook, who thought he was damn good at providing a tasty meal, was angry. During the night the cook loaded his baggage and left. When the hungry jacks woke in the morning, with their bellies grumbling, they found the table was set with bales of hay.

The old horse with the little sleigh must have been a welcome sight to the logging crews. Hot food on a cold winter day can pick up the spirits, not to mention giving you the fuel needed to keep an axe swinging.

The photographer arranged the loggers for the photo of this noon meal in the woods, but what is interesting to me is the background. There is not a living tree behind this logging crew, giving a clear insight into what was left for future generations, and giving credibility to the phrase, "cut and run."

Feeding 40, 50, or even 150 men three times a day must have been a formidable task. There wasn't a corner grocery available, and a good cook had to have his supplies put in well in advance of the season. He also had to be able to produce delicious meals that were filling and fit into the budget allowed by the owners of the operation. He usually had to work with young boys or old men, serving as bull cooks, who were responsible for keeping the wood boxes full, and he had to know how to fire the wood-burning cook stoves with the right wood to produce the correct heat for each recipe.

A good cook could make beans, salt pork, and rutabagas taste superb, especially if it was washed down with hot coffee followed by plenty of pie or donuts. Talking wasn't allowed in a cook shack. The cook didn't want anybody to linger over coffee and chitchat. He had to start preparing for breakfast, where pancakes were probably the most common fare. Bacon, side pork, or pork sausage was often served with the meal. It should also be noted that as the decades of the early logging era passed, the camps and the food improved, with the exception of the tail end of the era during the Great Depression.

> Harold Pichotta, Wabeno, Wisconsin, was the son of John Pichotta, a camp cook for Menominee Bay Shore, who not only cooked in the camps, but cooked and served food on the river drives held on the Pike and Menominee Rivers in Upper Michigan. John moved to Wabeno, Wisconsin when Bay Shore moved there and switched their operations to milling hardwood. John opened a hotel and continued to cook, but the bean-hole beans became a part of family tradition. Once a year, the descendants get together for beans cooked in the ground. According to the Pichotta family, there is a big "aaaahh" from the crowd when the beans are opened and the aroma hits the air. I'm sure the river pigs felt the same way.

All logging camps served plenty of pastry and pies. It was tasty and filling, and what was important to the owners was the cost. Pastry was cheaper than meat, the ingredients were easy to store, and the men liked to eat all the pie they could hold. Most cook shacks had a root cellar under the floor to keep things like potatoes, apples, raisins, and smoked meat. Many old logging camps sites are still visible by the foundations, which were usually just dirt banked around the building. The rectangle with the depression in it was probably the cook shack.

Camp life was loved by some and endured by others. Wooden bunks lined the walls, and a barrel stove or cast iron box heater usually provided the heat. The deacon's bench was just a plank near the stove or sometimes around the room using the bunks as part of the support. In the short evenings after supper, the jacks sat on the bench and told stories, worked on their gear, or sometimes played music if they were so inclined. There usually weren't many windows in the bunkhouse, but most had skylights that could be opened to cool off the place if the stove was over-fired, and it was also a way to get the tobacco smoke out and the rich aroma of wet wool drying near the stove. Sunday was a day off, and often it was spent washing clothing and bedding in the never-ending battle with lice and bedbugs. These were usually losing battles, and a good long, hot bath was needed in the spring, along with a new wardrobe.

This is an early camp on the Oconto River drainage, probably near Townsend or Lakewood, Wisconsin. The log construction usually dates camps to the early days of pine logging, when lumber wasn't available upriver. The bull cook has done his job, evident by the nice pile of cook stove wood piled behind the men. Cook stoves had small fireboxes and required smaller pieces of wood. The other reason for small wood was temperature regulation of the cook stove. Small pieces of firewood meant the cook could regulate the temperature in smaller increments and he didn't have to wait as long for heat to build or the stove to cool off.

Ain't it cozy! The bunkhouse in a logging camp must have given off a variety of odors. Wet wool, stale sweat, tobacco smoke, all added to the ambience. The surroundings were crude, but when a man was tired enough, the straw-filled bed felt good. The stories in the evening were entertaining too, and what student of logging history wouldn't like to go back in time and spend an evening listening to those yarns.

Lumberjacks wore wool in the winter, switching to overalls and flannel shirts in the spring. Stylish clothes had no place in the life of a man who avoided polite society. Footwear changed with the seasons. Early camp photos show men wearing low rubbers with several pairs of heavy wool socks. Some of the old-timers said they worked well enough and your feet stayed warm if you kept moving. As the years went by, rubber bottom pack boots became the standard winter footgear. L. L. Bean of Maine was famous for their rubber bottom packs, and they can still be purchased from Bean, but you can now opt for insulated bottoms, a feature not offered back then. Summer logging was frowned on by most of the lumberjacks, but during the log drives and during any summer work, leather boots with corks or nails in the soles were used. Staying on a slippery log would have been impossible without corks, and they were an absolute necessity in a good saloon fight. It was easy to see who had been on the floor during a saloon brawl, because the scars from corked logging boots were distinctive. Even the saloon floors took punishment from corked boots. Until recent times there were still some old-time saloons scattered around, usually in small, rural towns, that still had a splintered and punctured floor.

> Most of these old saloons are no more, but I must admit that it added a sense of history to beer drinking when the back bar was ornate and made of heavy, dark hardwood and the floor still had the scars of the old lumberjack's corked boots. I recall quitting a logging job near L'anse, Michigan and spending a rather "foggy" afternoon in one of these saloons in Amasa, Michigan. The old building had a tilt to it and a smell all its own. The floor was punctured with cork marks to all the edges, the beer was cold, and we too were lumberjacks after a fashion. Many years later, a carpenter from Wisconsin Dells showed me a coffee table made from a saloon floor that he salvaged in Amasa, Michigan. It brought back memories and put a lump in my throat that the old bar room floor could have such an ignominious end as to become part of a coffee table. I'm sure some old Michigan jacks rolled over in their graves when that happened!

This shot of a bunk house looks like the photographer did a little arranging. With all the musical instruments hanging about, you are left with the impression that the Milwaukee Symphony Orchestra sawed logs in the winter. Actually, many people played music in these days. If you liked music you learned to make your own. As you can see, there isn't a CD player or set of headphones anywhere in the photo.

The old bunkhouse is empty. The photo might have been taken after the men had left for the spring breakup. The bunks are muzzleloaders; they were entered at the end. Notice the blankets hung for privacy on the two bunks in the rear.

In 1948, an old logger and river man from Shawano, Wisconsin wrote down a brief autobiography of his life. His working days were spent mostly with the Keshena Improvement Company on the Wolf River and its tributaries. J. L. Whitehouse said that the camps he worked out of were well supplied and "….furnished with better and greater variety of food than the ordinary farms or city homes. He continued, "I never worked in a camp where I didn't get more to eat than I did at home, and I never starved or even went hungry at home." Whitehouse wrote the following about driving on the river:

> "Until we get below Shawano we live in tents. In every drive there are two crews. The rear crew of 30 or 40 men and the jam crew of about 25. They camp about six miles apart and are under the same boss, but have a foreman for each crew. There is a four-horse team kept busy hauling supplies from Shawano and moving them every week or ten days. The food is about the same as in the woods. Meat is salt pork, beef and smoked ham, the latter being used more than all others combined. When the walking boss comes up from town about once a week, he brings a day's supply of beefsteak and pork links for each crew. We also have an occasional feed of eggs. Pastry is used plentifully as it is the cheapest food to feed a crew. The walking boss has a driving team and drives from rear to front to keep things in running order."

The drives on the Wolf River were much the same as drives on other rivers throughout the Lake States. All provided dangerous work, and all brought in a broad spectrum of people who added color and legend to the area long after they were gone. The Wolf River was different than many of the rivers that handled drives, as there was a road that paralleled the Wolf from Green Bay in the south to the headwaters at Pine Lake, near Hiles, Wisconsin, and beyond to the Keweenaw Peninsula in Michigan's Upper Peninsula. The road was built by the federal government, and started during the Civil War because of a fear that Canada would help the Confederacy. That, of course, didn't happen, and if it had, the road wouldn't have been much help, because construction wasn't completed until a half dozen years after Lee gave up at Appomattox. But the road did make it easier for lumbermen to find timber stands, and businessmen, like saloonkeepers, were able to offer their wares upstream. River men on the Wolf, as the years went by, were able to belly up to a bar at any number of stopping places along the river. To be sure, the boss wouldn't approve, and often there wasn't time during a drive to sample the whiskey, but it did happen! A local legend that is mentioned in the Whitehouse autobiography occurred at a place called the Log Cabins. The Log Cabins "matured" from a roadhouse to a fishing camp and resort, and is now a home.

Sunday, the only day the jacks had a day off unless there was a blizzard, was often spent taking care of your personal equipment, and that included clothing. This gent is washing his clothes and running them through a portable wringer.

During the days of the river drives, The Log Cabins was still a place where a man could get a meal, some drinks and bed. Another saloonkeeper, upstream at Lily, was hauling a barrel of whiskey back home and decided to spend the night at the Log Cabins. George Trusdel had to stock his joint, and he turned down the requests of river men who wanted him to tap the barrel on the back of the wagon. They persisted, and George sat on the barrel all night with a Winchester across his knees. An enterprising river pig named Bogue Dickey borrowed a wash tub from the proprietor of the Log Cabins, and while his friends kept George busy and provided background noise, Bogue carefully drilled a hole through the wagon bed into the whiskey barrel and drained the hootch into the tub. George guarded an empty barrel all night, and the jacks had a fine time, with whiskey hid behind every stump and tree, it was a cheap drunk and a great joke. We don't know how George felt about the affair, however.

And so it was, the rivers of the Lake States carried the logs that helped to build the nation. The easy logging along the rivers was limited. As the pine disappeared into the mills, it became harder and harder to find good stands that were close enough to a river big enough to carry the load. Logging in the Lake States was far from over, as it is not over yet, but a different type of transportation was needed. Railroads were the answer.

Visitors probably didn't come to logging camps very often, but when they did, the best food and the best manners were hauled out. It was a logging camp tradition that people shouldn't leave hungry, and when talking to people who visited camps when they were kids, all testify that they were fed molasses cookies, raisin pie or pastries. All say they were excellent. The visitors in this photo were very likely given the same treatment. Notice the deerskin robes in the sleigh.

The jam crew was camping at the Log Cabins on the Wolf sometime in the 1890's, when this picture was shot. The Log Cabins was a stopping place for travelers on the Military Road, later was a fishing and hunting resort, and still exists today as a residence.

Chapter 2
Logging and the Railroads

The railroads made huge differences in the logging business in the Lake States. The biggest change was probably in the infrastructure. The value added to the logs could now be closer to the logging site. Sawmills could move their operations upstream, and the mouths of the rivers played a lesser part in the sawmill business. The trains could now carry lumber directly to Chicago, St. Louis and every other major city. The other big change in the industry was the ability to utilize the hardwood stands that, because of their greater density, couldn't be floated downstream to a mill.

Blackwell, Wisconsin was a sawdust town carved out of the woods. A spur was run into the new town to carry out the lumber from the Flanner sawmill. Most of the buildings in town were built by the Flanners, including the store, the boarding house, post office and all the homes.

Part of the Flanner mill shows a neat building with a sawdust burner (the silo shaped structure behind the mill). The Flanner brothers ran a nice mill and town, according to people who lived and worked there when they were in business. Unfortunately, the Flanner boys were one of the first casualties of the Great Depression. The story is that they liked to travel to the big cities and had a talent for picking "fast women and slow horses." Whatever the reason, the company went out of business, and Blackwell is a bedroom community today, surrounded by the Nicolet National Forest.

Logging sleighs didn't carry as many logs when hauling hardwood logs, because of the increased density of hardwood over pine.

Logging towns knew where the bread and butter came from, and if Keith & Hiles needed the main street of Crandon for a sleigh road, so be it. This load of hardwood was about 8 blocks from the mill on Clear Lake, and Harry was the son of M.D. Keith one of the owners of the mill. Actually, in the days before cars, there wouldn't be much traffic on the street during the winter, and the activity of the logging sleighs was welcome in keeping the street free from deep snows.

A load of logs is delivered to the Keith & Hiles mill. Not all logs were cut by company sawyers. Mills would buy logs from homesteaders or independent jobbers. The scaler is on the right. This is a big load of hardwood, and probably didn't come a great distance to get to the mill.

Loading pine with a single line. Small jobbers couldn't always afford the equipment or the horses, and they loaded logs with a single line. A chain was wrapped around the middle or balance point of the log and the horses pulled the chain from the other side of the sleigh up the skids. Men with canthooks tried to keep the ends of the log level, until it was over the load.

Loading with gin-pole loader in the woods. The men with the ropes in their hands steady the log as it is raised. The ropes are fastened to pup hooks, which stayed in the ends of the logs as long as there was pressure from the weight of the log. Sometimes they came lose, and everyone scrambled out of the way, hopefully.

The railroads spawned countless "sawdust towns" along the tracks, as entrepreneurs set up a sawmill and built a community. As the track lengthened, the number of towns grew. Often they were small communities with a few hundred people. As a rule, the town lasted as long as the timber stand feeding the mill. Other communities struggled along after the mill closed, with the logging industry gearing itself up to supply pulpwood to the paper mills that were being built in the Lake States. Tourism became an economic factor that kept some communities operating in the post-mill days, and some small sawmills turned to smaller logs and continue to operate yet today.

These tent dwellers were employed by the Keith & Hiles Company to make new grades and lay track for the upcoming logging season.

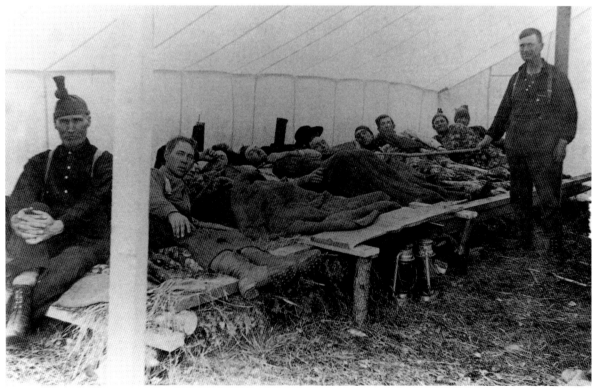

The same grading crew in the previous photo posed in their straw bunks. Imagine the snoring in that tent! Notice the corked boots adorning the feet of the man second from the left. They gave great traction on slippery logs and wet ground, and what a weapon in a brawl!

"So here's the deal boys. You shovel gravel on the flat cars for 12 hours. When we get one full, we'll pull it out and give you a fresh one to fill. The pay is a buck a day and your grub. You sleep in the tent on loose straw. If you don't like shovelin' gravel on the car, you can shovel it off the car at the fill."

Visitors at the camps on Sunday became more common in the railroad days. The rail line extended to the camp, and the communities were nearer to the woods than during the pine-logging era. This old shot is interesting in a number of ways. The remains of the forest in the background give a look at what much of the northern Lake States looked like after the early logging days. The gentleman with the Winchester was probably hunting deer for the camp's fresh meat supply. The tripod on the left with the wooden gambrel was the butchering spot for the deer and the camp's pig herd.

The late Al Krause, a friend who was born in Gillette, Wisconsin, spent a winter or two working for the Flanner Lumber Company, in Blackwell, Wisconsin. Al lived in the boarding house and loved to talk of the various characters who had moved north to find their fortunes in the newly opened logging territory. Al said the food was fine, but the facilities left a little to be desired. The bathroom facility was a lean-to built on the back of the boarding house, and the toilet was a pole suspended over a small stream. He said you could watch trout in the stream while performing your toilet!

The railroads and the buildup of small communities along the tracks also caused the development of another class of people who worked in the camps and mills, and moved to another location whenever they felt the grass was greener somewhere else. Company housing and company stores, along with credit from the lumber company, meant that families could set up housekeeping quite easily in another community. Often, workers were paid in script that was redeemable only at the company store. A family might not have much cash, but the basics were supplied. While this wasn't exactly a nomadic existence, it did mean that the labor force was much more mobile than in previous years.

A relative of mine, Uncle George Monty, was a woods boss who followed the Page & Landeck mill from New London, Wisconsin to Crandon, Wisconsin in 1901. My grandfather would come north from the farm to work the winters in Uncle George's camp, until he decided to stay in the 1920's, and that is why I live where I do today. Uncle George, however, developed the wanderlust of the times. His son, George, Jr., told me that after leaving Crandon, they went to the western part of Wisconsin, to Weyerhaeuser, for a time. The wanderlust had set in and Uncle George moved his family in a large circle from sawdust town to sawdust town. George, Jr. told me that he graduated from high school in Escanaba, Michigan, 100 miles east of Crandon, after attending school in eighteen different communities!

Another group of people came to the backwoods because of the railroads. The homesteader. Railroads greatly facilitated the building of farms out of former forestlands. Every lumber company also became a land company, and as soon as the trees were cut, cheap land was offered to would-be farmers who wanted to escape the cities and dirty factory jobs for fresh air and independence on their own piece of land.

Two types of railroads came into existence in the Lake States. The big commercial carriers like the Soo Line and the Chicago & Northwestern relied heavily on the timber industry for freight revenues and expansions of their lines. The federal government also gave land, every other section in some areas, to the railroads as an incentive for expansion of the rail. This meant that railroads could sell land and standing timber to companies that would soon be their best customers. And, the logging companies started using their own railroads to get logs from the woods to the mill, or, in some instances to haul lumber from the mill to the commercial carrier.

The Keith & Hiles Lumber Company operated out of Crandon, Wisconsin. The company started life as the Page & Landeck Lumber Company. They bought land around the Crandon area in 1892, but they were unable to utilize the timber resource until 1901, when the C & N W built a spur into Crandon from Pelican Lake. The company changed its name to Keith & Hiles when Franklin Pierce Hiles sold his mill and his town at Hiles and bought out the Landecks. Their Shay engine hauled many millions of board feet to the mill on Clear Lake. The photographer caught the logs rolling down the ways into the hot pond. The cars were equipped with trip stakes that could be released, letting the load roll off.

50

A view of the Keith & Hiles hot pond with the mill out of sight on the right-hand side of the photo shows how the logs were handled and sorted and brought to the live chain that brought the logs into the second story of the mill for sawing. A hot pond performed several functions. The water cleaned the sand and other grit off the logs, keeping the saw sharper longer. Because the hot pond portion of the lake was kept ice-free during the winter by venting waste steam into the water, the frozen timber thawed out, making the wood easier to saw, and it was a handy sorting device for different species of logs. Some of the heavier hardwoods, if they were left in the water too long, would go to the bottom, where, undoubtedly, some still reside.

This is a nice view of the Keith & Hiles Shay engine. The mechanically inclined can see how the drive train worked off of the three steam engine cylinders directly in front of the cab.

The Keith family had this photo of Louis Christensen in their files. Louie followed the company north from New London in 1901, and was a good camp cook. He finished out his days in Crandon. A versatile man, he took up gunsmithing in his later years, and put together some rifles that were tack drivers. Men like Louie were a big asset to a company that needed men it could depend on.

Logging railroads were usually short run affairs, sometimes held together by baling wire and hope. They serviced a sawmill from a spider web of spurs, often feeding into a main line that ran to the mill. The spurs were abandoned as soon as the timber was cut, and another line was pushed into an uncut stand. The crews that laid the rail worked through the summer months to prepare the roadbed and lay the ties and rails salvaged from the abandoned spurs. Allowances had to be made for steep grades, and when the hill was too big for the engine to climb or for the track crews to cut through, the line grew longer as it wound around the hills.

A new siding has been laid in along the main line. At first glance it is an uninteresting scene, but at closer inspection, you can see that the ties are hand-hewed. It took a skilled axe man to hew ties using a broadax. The tree was sawed down and two opposite sides were flattened with the axe bringing the tree to the appropriate width. The tree was then sawed to length and taken out of the woods. Men who worked handling ties could tell terrible stories of backbreaking work with low pay. When ties were transported by rail, they were hand-loaded into boxcars, with the men spending the entire day carrying one heavy oak or maple tie into the boxcar. The same horror stories came from the men who later hauled ties on trucks to the railhead.

Life wasn't always easy on the railroad! A derailment could upset schedules, woods bosses, head sawyers, the owners, and really upset the men who had to use cant hooks, bullwork, and awkwardness to clean up the mess!

Working on the logging railroad was a coveted job. While sawyers and skidders toiled in the cold and deep snow, staying in camps, the railroaders worked by a boiler that could get hot enough to make your overalls smoke. They were looked up to by people in the community as skilled workmen, who not only make a contraption like a steam locomotive operate, but could also repair it when it broke down. The engineer and fireman could help themselves to some fine table fare while in camp, but usually slept at home with the wife, not in a smelly bunkhouse. It was a romantic job with a steam whistle to toot when you passed your friends or a group of fascinated kids.

Most logging railroads were equipped with small steam engines, and some, but not all were narrow gauge, meaning that the tracks were laid closer together. This meant that the rolling stock from a narrow gauge couldn't operate on a standard gauge tract and vice-versa, necessitating unloading and reloading of logs or lumber from the narrow gauge to the standard gauge commercial carrier.

The crew of the Rogers Lumber Co. sawmill poses with the company engine. Rogers Lumber created the Town of Nashville, Wisconsin out of the wilderness and connected with the C & N W railroad nearby. Around 1900, they built a sawmill, company store, brought in some people, and soon there was a church and a schoolhouse. In 1914, the mill burned down and wasn't rebuilt. The village slowly broke up and many of the buildings eventually fell down, including the homes of the Rogers family.

LOADING CARS NASHVILLE WIS.

The Rogers Lumber Co. crew is loading cars with an a-frame jammer. Note the logging camp in the background. Supplying the camp was a lot easier in the railroad days, and the men were able to eat more fresh meat and vegetables.

This picture of the Rogers locomotive is for the railroad buffs. What is also interesting is the size of logs they were hauling. In the early days of logging, the small trees were passed by, but later on, especially when logging hardwoods, the size of the trees harvested became smaller, perhaps, because they could see the end of the easy timber sales.

Below:
A dray load of cedar posts heads to market. In many of these old photos you will spot piles of small logs too small to run through a big mill, and that will be the cedar post pile. Cedar posts were a mainstay of most of the logging companies. Cedar resists rot for decades, and was used in the Lake States and the western prairie states for stringing barbed wire. Many of the posts logged in the Lake States were used in mining operations in Minnesota, Michigan and Wisconsin. The posts were used in the mines to prevent cave-ins. Cedar was also used for shingles. The bigger posts would be cut to shingle length and run through a cedar saw, which was a circle saw running horizontally. A cedar sawyer had to run his fingers dangerously close to the saw to get the final shingles from a block, and many of these guys could be picked out of a crowd by their lack of digits.

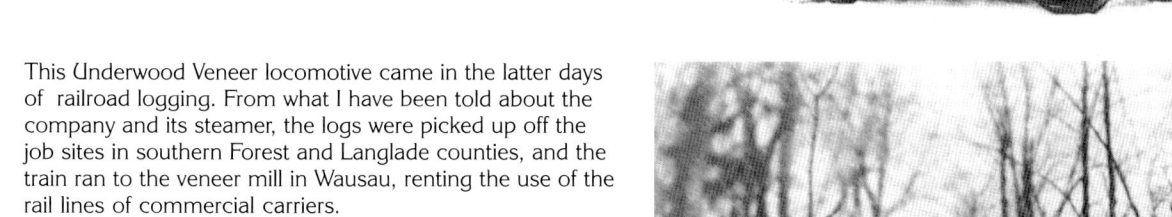

This Underwood Veneer locomotive came in the latter days of railroad logging. From what I have been told about the company and its steamer, the logs were picked up off the job sites in southern Forest and Langlade counties, and the train ran to the veneer mill in Wausau, renting the use of the rail lines of commercial carriers.

A popular steam engine for logging was made in Lima, Ohio. It was the Shay engine often called a "Limey" after its place of birth. The Shay wasn't equipped with the external horizontal drive arms to supply power to the wheels. Instead, power ran from the steam engine through vertical push rods that transferred power to the drive wheels through a series of bevel gears and shafts. The result was a tough little engine that was geared low, and had plenty of power, though not much speed. The advantage was that power could be supplied to the drivers slowly and without jerking movements, keeping the wheels from breaking traction. The little limey engines could climb steeper hills and were more suited to the job of logging. Railroad grades didn't have to be as flat, so the time and expense of cutting grades through hills was greatly reduced.

Cook shanties of the railroad era began to look modern. With a railroad track running past the front door and at least one train pulling up every day, it became easier to supply a kitchen with all the necessary utensils and maybe some that weren't absolutely necessary. In spite of better utensils and cookware, their job was the same as the cooks in the pine-logging era. Plenty of grub, hot coffee, and good pastries and pies. Every cook still maintained silence in the cook shanty. The men weren't allowed to tell tales or light a pipe at the table. That was for the deacon's bench in the bunkhouse, because as soon as the tables were emptied the dishes, cups, pots and pans had to be washed and ready for breakfast.

Right:

How I spent my summer vacation...... In these days, leather wasn't the luxury item it is today. The world depended on leather for countless daily uses, and that leather had to be tanned. Tannic acid was extracted from hemlock bark and used in the tanning process. This rugged individual is spending his summer peeling hemlock. Hemlock peels in the summer when the sap is up in the tree. Most logging companies didn't require the services of many lumberjacks during the summer months, and rather than do farm work, loathed by most good jacks, this gent is in the tan bark business. Note the long handle of his hemlock spud in front of the one-man saw. This was a highly wasteful business. Hemlock wasn't a species preferred for lumber, so often the tree was felled and peeled and the wood left to rot. The bark was tied in bundles and hauled by wagon to a collection point, usually a rail siding.

The bark house will keep out a few bugs and some of the rain. The famed architect, Frank Lloyd Wright, preached that houses should blend with the landscape. Mr. Wright could have learned from this lumberjack.

As the railroad companies expanded their lines and towns sprung up, there was jubilation. Usually, the good feelings were short-lived. As the freight rates climbed, the railroad companies soon came to be looked upon as unfair and were accused of price gouging. Railroads didn't immediately replace river drives. The evolution was slow, and both methods of getting pine timber to market were used at the same time. There was an overlap, and some of the pine lumber companies went back to river drives when freight rates made the rivers more profitable, but trains were here to stay, and eventually totally replaced the rivers as a means of transport for logs.

The woods boss looks over the cut before the men skid out the logs. This may not be the woods boss, but it probably is. He appears to be a man in his fifties, and this is about 1910. If he has been a woodsman for his entire working life, he probably started in a logging camp in the 1870's as a chore boy or cookee in the cook shack. He would have grown with the logging business into other jobs that enabled him to understand logs, logging, how to handle rough men (sometimes with fists), and bosses who wanted more and more production at less cost. He was mentally and physically tough and had earned the respect of the men he controlled and the men he worked for. By the way, a log cutter today would be fired for leaving stumps that high and wasting wood!

The Page & Landeck Lumber Co. of Crandon, Wisconsin waited for ten years to erect their mill. They purchased their timberlands in 1891. The Chicago & Northwestern finally ran a spur line from Pelican Lake to Crandon in 1901, facilitating the construction of a big mill that produced 80,000 board feet a day in two ten hour shifts. There was jubilation and celebration, not too mention speculation. Crandon tripled in population in a few years and it was boom times, but by 1906, the honeymoon was over. A company was formed to build a rail line from Crandon, north to a junction with the Soo Line near North Crandon, now Argonne. The consensus was that the C & N W was too pricey and the service was poor. The Wisconsin & Northern was eventually constructed from Argonne to Menasha, Wisconsin, and was bought by the Soo Line.

Below:
Another product supplied by the northern forests of the Lake States was bridge pilings. Select, straight trees of the same diameter were used to provide footings for bridges and other structures that were heavy. A pile driver would pound these trees into the ground and the bridge was built on the butts. The lack of oxygen in the soil kept the pilings from rotting, and it is likely the trees in this photo are still in great condition, somewhere, deep in the ground.

Opposite page:

One piece of infrastructure that moved into the woods with the railroads was the saloon. In the early pine-logging era, the nearest saloon would be many miles from the camps, a great help, and I'm sure, to the productivity of the camp. Saloons sprung up along railroad tracks, often it was the first building built. The saloon was the meeting place and served multiple functions in a new community. The particular watering hole was in North Crandon, now Argonne, in the 1890's. This time period was the transition point between the pine logging era and the railroad era. The Minneapolis, St. Paul & Sault St. Marie Railroad came through in 1887, creating the community and giving rail service to the immediate area. While this saloon was operating, log drives were being made on the Wolf, the Pine and Peshtigo, very near North Crandon, but if you build a railroad, they will come, so logs were loaded on railcars on the Soo Line, as it came to be called, and sent to mills in Green Bay and other cities.

Right:

Railroads made it easier for everyone to go share in the boom times in the big woods. As logging and sawmill communities grew, the need for temporary shelter expanded. The hotel business grew to accommodate the needs of the many travelers. The Grand Plank Hotel, in Wabeno, Wisconsin, was across a muddy road from the C & N W depot. Planks were laid across the road so the newly arrived didn't have to wade the mud, and the hotel, I'm sure, was named with tongue-in-cheek. The name was put on the tin-covered three-story building, and so it stands today, closed finally in the year 2000. Ceil Gayhart, the current owner, finally retired, but the old building served for nearly a century as a hotel, boarding house and finally a tavern. Guests were served whatever was available, and that might include venison, bear meat, suckers from the nearby creek, and wild berries in season. Complaints were few, and if you did, too bad, that is what was served, take it or leave it.

The Michigan House was a welcome sight to weary travelers near Armstrong Creek, Wisconsin. Mrs. Ball and a daughter operated the establishment. I was shown the location a few years back and walked around in the cellar, which was sprouting large trees, but it was easy to see how root cellars kept vegetables and liquid refreshments cool in the summer and kept them from freezing in the winter. It was located on one of the early roads that would take the traveler to the Iron Mountain, Michigan area, and was a short walk from the Soo Line Railroad.

The Eagle Saloon was one of the first saloons in the new Town of Wabeno, back in 1897. The proprietor, John Mallow, was a hard-core saloon man, and like many saloon keepers back then, involved in politics. A local legend tells of John Mallow watching one of the many homesteaders go by with an ox cart full of staples like flour and corn meal. He shook his head in disgust and said, "Look at that sonovabitch, all those groceries and I'll bet he doesn't have a drop of whiskey in the house."

When the thirsty sawmill worker or lumberjack entered the Eagle Saloon or one of its many counterparts, this is what he saw. Spittoons were a fixture of all "decent" saloons. As can be seen from the tobacco cuds on the floor, none of the customers could win a spitting contest. Still, the saloon was warm and friendly, there was often a free lunch, and was a male refuge from the outside world.

In Crandon, Wisconsin, the new Northwestern Hotel was a block away from the C & N W depot. No saloon in the Northwestern, or anywhere in Crandon in those days. Sam Shaw, the founding father, along with Women's Christian Temperance Union member, Louise Shaw, Sam's wife didn't allow hard drinks in their town. A short walk to the outskirts could get the thirsty traveler a bottle, and the inconvenience was not rectified until after prohibition was enacted and abolished.

Wiley Ferguson operated the Northwestern Hotel, and if you had just walked in from the depot, you would have seen this sight.

It is always amazing to me how fast sawdust towns were built. The Hotel Wabeno was built between 1897 and 1900, along with the two businesses next door, numerous saloons and stores, houses, a school and churches. And, it was done without an array of power tools. There may not have been a lot of square corners in these towns, but the buildings kept off the rain and snow, and gave the traveler a warm bed and hot breakfast.

Decking logs was a never-ending job. The sheer amount of timber taken off the land meant that the logs had to be piled high just to fit the space available. After sleigh hauling the logs to a railhead, the logs would have to be decked again, and might also have to be decked at the mill. Of course, whenever possible, loggers "hot-logged," skidding to a sleigh where the logs were loaded and removed. Small operators could get away with this, but the big companies that were handling massive amounts of logs usually had to deck. This is called a gin-pole loader or jammer. A certain amount of flexibility in the arm made this loader more handy than the old a-frame jammers.

Types of jammers varied with the men who designed and built them. This unit operated in the mill yard and would have been too large and cumbersome for the woods. The boom was adjustable for longer reach, and the logs that form the base are long so the unit won't tip. It was probably powered by a team of horses.

Bob Hewitt's decking crew near Crandon, Wisconsin shortly after 1900, in the fall of the year before the snows. Nine men are involved. Maybe some crowded into the photo who weren't in the crew, but it is a good indicator of how labor-intensive early logging was, and this photo also shows the inherent danger in stacking heavy cylindrical objects. If one of those logs had slipped out of place, tons of wood rolled over and smashed anything in the way.

The logs are waiting under the snow. The woods boss is looking over the next sleigh haul, and soon these quiet woods will be ringing with shouts and cursing, snorting horses and the sounds of wooden runners on ice trails.

A decking crew takes time out for the camera atop
tens of thousands of board feet of hardwood logs.

The Page & Landeck jammer was loading logs on the C & N W tracks when these photos were taken in 1907, near Crandon, Wisconsin. These steam-powered loaders were called "slide-ass jammers." The loader sat on a flat car and loaded the next car. When that car was loaded a cable on a winch drum pulled or slid the jammer to the next car, where the process was repeated, thence the name. Notice the man with the "cheat stick." He's the scaler, and the stick is a scale rule. Scale rules have always been cursed by loggers, and they probably always will. The scale rule is held against the small end of the log and the amount of board feet is read from the column for the length of the log. You could never convince a logger that scale rules weren't invented to give him a thorough screwing, and he might be right. *(Shown above and opposite page)*

Steam power in the woods was possible with railroads, and steam powered loaders facilitated quicker handling of logs. These pine logs are being loaded on C & N W rail cars for the trip to the Holt mill at Oconto, Wisconsin.

The Lake States are crisscrossed with old railroad grades that are sometimes utilized as roads today. Indeed, the grades weren't used as railroads very long. As soon as an area was logged off, the rails, ties, plates and spikes were pulled and moved to the next logging chance. This recycling was done until the holdings of the company were exhausted. Usually those tough little limey engines were sold to another logging company, or sometimes a mining company. Many of them ended their useful career in Colorado, Arizona or even in Mexico or South America. A few were left to rust in the woods, and may have slowly turned back into soil if it weren't for Hirohito and Hitler. World War II provided the market for scrap metal that turned many a steam engine into Liberty ships and Sherman tanks.

Another piece of machinery used to move logs to the mill was the Phoenix Log Hauler. The Phoenix is an adaptation of the Lombard that was developed and tried in Maine. The hills of Maine proved too much for the Lombard. Phoenix Steel, located in Eau Claire, Wisconsin, acquired rights to the Lombard and developed the Phoenix. In many situations the Phoenix was the right answer, and proved more profitable than a logging railroad.

The Phoenix was a cross between the crawler tractor and the railroad steam engine. Like horses and sleighs, it required a sleigh road with ruts cut for the runners to ride in. Unlike horses, the Phoenix could pull great strings of sleighs loaded with heavy hardwood logs, much like a logging train could pull loaded rail cars. The engine with the long string of sleighs curving around the hills resembled a giant serpent, and the Phoenix acquired the name "snow snake." The advantage of the snow snake was that the grade didn't require as much leveling as a train, and the expense of buying rails, ties and spikes was avoided. Like a train, the Phoenix needed an engineer who could also pull the job of fireman. Steering was done with skis mounted in the front, and a steersman who braved the cold out in front of the warm boiler. The Phoenix was used throughout the Lake States where the terrain would permit, and was also used in places like Finland and Russia. Sometimes, the Phoenix was used in the wheat harvests on the Great Plains, but their most useful application was in the woods.

G. W. Jones Lumber Co. out of Appleton, Wisconsin, used the Phoenix extensively in their Wabeno, Wisconsin operations. Many of the older residents can recall quite clearly the Phoenix chugging into town with its long string of log sleighs groaning on the ice road. Jones wore out one Phoenix, and possibly (the records are incomplete) two more, supplying logs to the Wabeno mill. Menominee Bay Shore, also located in Wabeno, opted for a train to keep the logs coming to the

An elderly lady who grew up in the woods in the early 1900's, my maternal grandmother, told how her father plunged into the sawmill and logging business. Buying a small sawmill on the Pine River, in Florence County, Wisconsin, he promptly went broke. The hardwood market in 1911 was shaky. Still, Great Grandfather persevered. He bought a Case steam engine that he rented out to small sawmills and also powered a threshing machine during harvest season. The Case could move, albeit slowly, from job site to job site. He was a man who jumped from venture to venture, but a few years later, he made the final payment while renting the steamer out to a sawmill. He immediately looked for other opportunities and seems to have put the Case out of his mind entirely, and he never went back to get it. The Case sat in the brush for close to two decades and was unceremoniously scrapped out to help the war effort in the 1940's. While this seems, and is, rather erratic behavior, it was not that unusual in a time when the Northwoods was filled with people bursting with entrepreneurial spirit, looking for the break that would bring wealth.

The Phoenix hauling logs to the mill for the G. W. Jones Lbr. Co. in Wabeno, Wisconsin.

yard. It will probably be debated forever as to which method was best, but possibly, Bay Shore had rougher terrain and found that sleighing into a main rail line was the better way. Regardless, the Phoenix worked well for a great many logging and sawmill companies, and pictures of snow snakes show up in small museums throughout the Midwest.

The last Phoenix used in Wabeno is now owned by the community and is on display in the middle of town next to their logging museum. It has become the town symbol, and has been restored to working order. Wheels are substituted for the skis, and the old tractor is fired up and run at most town events. As far as is known, it is the last operating Phoenix in the world.

The G. W. Jones Phoenix negotiates a corner with a long string of sleighs.

The Phoenix Steel Company, Eau Claire, Wisconsin, knew how to advertise their product. They sent out photographers to bring back pictures of the Phoenix working, and the photos were used to sell the machine to other lumber companies. They sold and manufactured a lot of log haulers, but they also contributed to the photographic record of the era. This phenomenal photo of the Jones Phoenix was taken just outside Wabeno on the shores of Range Line Lake. The Phoenix was pulling in over 100,000 board feet of hardwood logs to the Jones sawmill about a mile from this spot!

76

It is said that the tracks of the Phoenix were the model used to build tanks during World War I. Whether this is true or not, tracks became a common sight in the woods during the 1920's. Brand names like Holt, Best, and Caterpillar became common during that time frame, and gasoline engines, not steam, powered them. The Holt Tractor Company produced various models designated by the ton. There was a Holt 2 Ton, 5 Ton and 10 Ton. These early tractors weren't equipped with blades or any hydraulics. The machines were used to pull sleighs loaded with logs to the mill or railhead, and often replaced the horse in pulling logs to the skid ways.

This is a great photo of the Menominee Bay Shore sawmill during it's peak years during the 1920's or slightly earlier. The mill was brought to Wabeno from Menominee, Michigan after the river drives ended on the Menominee River and its tributaries. The mill actually wasn't in Wabeno, but in a new town created next to Wabeno, called Soperton, in honor of the principal owners, the Soper family. They called themselves the "Big Red Mill," and put it on their stationery. For a time, the Bay Shore was the largest hardwood sawmill in the world, later to lose the title to the Connor Lumber and Land sawmill nine miles to the north. The tramway system was devised to help pile lumber. There were no dry kilns, and lumber was air dried in precise piles that were slanted to shed rain and melting snow. Because of the great volume of lumber on hand most of the time, the piles were high, allowing more space for lumber to dry in the mill yard. The tramways gave a nice surface for horses pulling carts laden with lumber to travel on, and also gave the lumber stackers a high spot from which to work. Another use for the tramways, according to older Wabeno residents, was courting. Couples used to make romantic walks around the tramways on weekends.

The hot pond at Menominee Bay Shore in the late 1920s or early '30s, judging by the automobile in the background. Bay Shore and Jones were both located on the North Branch of the Oconto River, side by side. The river is still a good trout fishing spot, and according to old-timers, was good fishing all the while the dams were in.

As the big timber was cut, smaller timber started to look better. This crew is working a hot pond, probably at the Bay Shore mill. The small stems could produce a couple of 1x4 inch boards, and Bay Shore was noted for utilization of its timber stands. The company existed until the late 30's, when the mill was split in half with one carriage or side sold and moved to the western side of Wisconsin. Bay Shore became Soperton Lumber Co., but not before they tried as many uses for small lumber as possible, including pre-fabricated cottages moved to your site. This was an attempt to cash in on the growing tourist industry and the development of lake frontage. Some of the cottages were sold, but they were a little ahead of their time, and World War II changed the market for everyone.

The Connor Lumber and Land Company, Laona, Wisconsin, was the biggest producer of hardwood lumber in the world for years. The company located on the Rat River before 1900, and built a town around the mill complex. And, they also were instrumental in the founding of Aurbundale, Wisconsin before coming to Laona. Connors produced hardwood flooring and hardwood juvenile furniture for decades, and marketed these products nationwide. According to company sources, the floor of the New York Stock Exchange and Madison Square Garden were planed and tongue and grooved in Laona. The Connor family's biggest contribution was probably in the acceptance and use of sustained yield forestry, which kept and is still keeping a healthy, productive forest around the town and the mill. The Connor Lumber and Land Co. was sold to Swiss investors in the 1970's, and in the 1980's they moved the entire operation, leaving Laona without a mill for the first time since it was founded. Fortunately, Gordon Connor stepped in and revived the mill, continuing the good forestry practices on the 50,000 or so acres they own, and is pouring out high-grade lumber as I type.

Charles W. Fish was a mover and shaker in the early logging days. He operated a sawmill and town at Elcho, Wisconsin, he bought the Keith & Hiles mill in Crandon, and he bought the Forster-Mueller holdings in Hiles, Wisconsin. After leaving Wisconsin, he went west and made some more money in the timber business. Fish liked to mold the communities he owned, and in Hiles, he painted everything that didn't move white. The company store in Hiles still exists in a cut-down form as a home, but I remember buying candy and ice cream in this building when it looked very much like this in the 1950's.

The boarding house at Hiles before Fish bought the business. Boarding houses were an essential institution in an industry that had people coming and going. Bachelors were housed in the boarding house, and the employer could be sure that he had a good place to sleep and three squares, keeping him in shape for hard physical labor.

Hiles must have been a terrible place to live if you liked the color green or red. The company houses, I have heard, virtually gleamed with white paint, and the electrical streetlights, powered by electricity from the mill, induced some wag to refer to this street as the "Great White Way." At 9:00 p.m. the generator was shut off, and the streetlights went out. Still, if it was monotonous, it was well kept, and many mill towns weren't. Some of these houses are still being used, but in a much modified form.

SCENE IN PARK
HILES WIS.

The same Hiles boarding house after Fish Lumber Co. made a few improvements. The fountain was a nice touch, but hard to maintain. Forty years ago or so, the remains of the fountain were scattered around the concrete base and the boarding house had long since ceased to exist. It was a good place for kids to sit and enjoy an ice cream bar and wonder what it looked like with the fountain operating, as I remember. Well, kids this is what it looked like!

The planing mill at Hiles. Like all mills of the era, horsepower and special carts moved the lumber around the yard. Boards were all handled by hand, and nobody had ever heard of a forklift.

Early crawler tractors weren't equipped with any operator protection. The driver of a tractor, especially when skidding logs, was exposed to any limbs or dead trees that fell over the tractor, and this was common. The heavy, cumbersome machines not only shook the trees they hit, but could shake the ground as well, causing dead standing timber to fall. A falling dead limb is called a "widow maker," and they earned their name when crawler tractors came into the woods. Today, nobody takes a piece of machinery into the woods without elaborate operator protection overhead and on the sides of the machine, but back then a lost man was just part of doing business in the woods.

The Cleereman-Jauquet Lumber Co. was originally located in Green Bay. They bought land in northern Wisconsin in the 1880's and set up winter camps to log their holdings. They were limited, however, to cutting timber about six miles from the Soo Line tracks, because the horse-drawn sleighs couldn't profitably haul logs beyond that distance. In 1911, the C & N W laid rail through Newald, a clearing in the woods that had a trading post/saloon to serve loggers and homesteaders. They built a modern steam-powered mill and moved their operations north. Never afraid of trying new technology, they bought a Ten Ton Holt and a Five Ton. This Ten Ton Holt hauled logs into the Cleereman/Jauquet mill in the 20's and 30's, and horse teams were also used.

Frances Cleereman told me about being a kid when the family bought these machines. Unlike horses, they didn't have to be fed and bedded down each night, and they didn't require any shoveling in the morning. This was in the days before anti-freeze, however, and the big machines did require water in the radiator. Frances said they built sheds with stoves that had to be fired each night, or the cooling systems had to be drained, and the big radiators refilled in the morning, a job he didn't enjoy.

The Ten Ton's kid brother, the Five Ton Holt, couldn't pull as much, but was used extensively in the Northwoods.

Enterprising lumbermen like the Cleeremans and the Jauquets needed good people behind them to accomplish what they did. One such man was Tom Kleve, the man on the right in this photo. Kleve was a runaway who found himself, as a young boy, in the village of Cavour, Wisconsin. Taken in by Charles Ross, a farmer, logger and politician, Kleve settled in and started learning about the new world he found himself in. He learned logging, saw milling and a lot about being a millwright. He found work with Cleereman-Jauquet, and became fast friends with Bill Cleereman, the man on the left in this photo of Tom Kleve's cabin. A look around the room shows the varied interests of Kleve, including taxidermy, judging by the stuffed squirrel and duck. He shot a Winchester autoloader rifle, either a .401 or .351, leaning against the table. He was obviously a very intelligent man who would have probably felt at home in Thomas Edison's lab, but he was a backwoodsman. Kleve's biggest contribution to the logging industry was that he trained Bill's boy, Frances Cleereman, the art of millwrighting. In the early '50s, Frances invented the first successful automated sawmill carriage and completely changed the sawmill industry. The Cleereman Sawmill Carriage is still being manufactured and improved today across the street from Bill Cleereman's house, in Newald, Wisconsin.

Pride in his product shows in this photo of pioneer lumberman Bill Cleereman. The high-grade shingles he is holding were the standard roofing material of the day.

The Ehlinger Brothers used a Holt Five Ton in their operations near Hollister, Wisconsin. They logged the hardwoods that were left in the pine era along the Wolf River corridor.

The hardwood-logging era in the Lake States finished off the great stands of virgin timber. Trains made it possible to log off the last great tracts, although, some nice stands did last for a while longer. With river logging, the timber had to be floatable, ruling out hardwoods. It also had to be within economical sleigh-haul distance of a stream big enough to dam and float timber, usually about six miles. The railroads were able to haul logs out of areas too far from a stream and gave lumber companies a chance to utilize hardwoods, but it was still a labor-intensive and expensive way to operate. If an area wasn't completely logged off in a season, and it was time for the rails to be moved, some stands were left. It just wasn't good economics to tie up the rail and the rolling stock for a small cut. Of course, these stands furnished logs for the next generation of loggers to cut and haul with trucks.

The Ehlingers had one of the fanciest icing rigs I have ever seen, equipped with a window! It is obvious that the internal combustion engine was working its way into the woods, as this ice wagon was equipped with a pump to eliminate hand dipping and frozen woolen pants.

This gent is cutting white spruce. While spruce makes fine lumber, my guess is that he is cutting pulpwood for Wisconsin's growing paper industry. Note the tree he has stuck his axe in. The blaze marks indicate a property boundary.

This Caterpillar 21 was operating on the George Palmer operation in the Blue Hills area near Ladysmith, Wisconsin. The crawler was used to skid logs to the road, however, and not pull sleighs to the mill. The Caterpillar Tractor Company was noted for making durable machinery, and some of these old crawlers ran for decades. Shortly after I had acquired this photo for my collection, I took a fishing trip to Canada and stayed in a rough-lumber lodge on an island. I walked out the back door, and there was a small sawmill and next to it a Caterpillar 18, an older model than this 21. I was amazed and was looking the old girl over when the owner walked up. I asked if the old Cat still ran, and he looked at me like I was a total idiot; "Of course it runs! How the hell do you think I skid the logs into the mill?"

The Palmer operation in the Blue Hills was also using trucks to haul their logs. The timber was still being loaded on the truck with an a-frame jammer, though, showing that technology took its time coming to the logging job.

The old Cat 21 could perform this job better than horses. Pulling a loaded truck up an icy hill would have been extremely difficult, if not impossible, for a team of horses.

Reading some of the many histories of logging in the Lake States, it would appear that logging and saw milling of any scale stopped about 1900, with small river drives continuing on some of the rivers until the early 1920's. While it may be true that some of the peak years of logging occurred during the pine era, and some colorful characters are from that time, the truth is that logging was a powerful economic force for several decades, with some companies logging by rail into the 1940's.

> *Back in the early 1970s, while logging a piece of hardwood timber up near the Michigan border, I looked over the cutting line and noticed long, moss-covered humps covering the forest floor. Putting down the saw, I walked over into several acres of pine logs that had been cut and never skidded out. The cuts were clearly visible in the moss. I was perplexed, until I gave it more thought. The Boom Lake Lumber Co. out of Rhinelander, Wisconsin, used their rail line, the Thunder Lake Railroad, to log off thousands of acres in that area. Evidently, spring breakup came a little early and the logs weren't skidded. The rail line had to be moved to another area, and the logs were left to enrich the forest soil.*

Logging hardwood was a different game, and different skills were required. In many ways it was more dangerous for the man in the woods. Hardwood trees are much heavier than pine, and they had a tendency to split when being sawed off the stump. Called a "barber chair," the tree would split in half and pivot on top of the remaining portion of the tree, which could be ten or even twenty feet over the heads of the lumberjacks. A barber chair usually happens suddenly and sounds like a rifle shot. Wide-eyed sawyers plunged through the snow to get away from the stump, because the tree usually fell off the stub. And, they had better run sideways from the tree, because as it pivoted the butt stuck out backwards the same distance as the height of the stub. Hardwood is heavier, and falling limbs weighed a considerable amount more than pine limbs. The widow makers worked like a dart, with the heavy end leading the way, and the small ends acting like the feathers. Widow makers caved in many fragile skulls, especially in the days before hardhats. Indeed, graves are being dug, yet today, for men who aren't careful enough when cutting hardwood logs.

This is an unusual photo of a load of logs, in that it is on the move. Most photos of sleighs loaded with logs are stationary and posed. Look closely, and you will see that there are four horse and two oxen hooked to this load, and the teamster is using his goad stick on the oxen. My guess is that the road is not frozen down, and the sleigh is pulling hard. It is either the beginning or the end of the logging season. Note the hand made sleigh on the right and the logging sleigh in the foreground.

The weight of the hardwoods changed harvesting methods and made logging more expensive. Animal power was still the predominant motive force in getting logs to the rail line, and heavier logs meant smaller loads on the sleighs. Throughout the early logging era, big sleigh loads of logs were photographed with the logging crews and office staff standing proudly by their handiwork. Often two horses were hooked to the sleigh, and usually, these two horses delivered the load to mill yard. However, it was always a downhill pull, and usually another team was hooked to the load to get it started. These loads make great pictures, but just weren't the normal way logs were hauled. Never, are these monster loads made up of hardwoods. If they were, the sleigh would probably collapse. For these reasons, machinery began to appear in the woods.

Houses aren't built from hardwoods, they are decorated with hardwoods. Hardwood floors became popular. Hardwood moldings, door and window trim became the style, even in average homes. The dense oak from the northern states also made excellent furniture, but hardwoods were also utilitarian. Hardwoods were used in framing machinery and in early automobile bodies.

The railroad line that sold the timberlands to the lumber company needed an inexhaustible supply of ties, and often became one of the lumber company's best customers. Whiskey barrels were needed to age whiskey, transport flour and other commodities. Fruit was transported in little hardwood boxes, and eventually, pallets became, and still are one of the principal means to move freight. As the supply grew, so did the market.

It was during the hardwood-logging era that waste became important with sawmills. Utilization became a buzzword as the end of the big tracts of timber came into sight. The width of a saw blade and the amount of sawdust it produced became an issue, and markets for small pieces of wood became important. A sawmill operator thought twice about throwing a narrow board into the firebox of the steam engine if he could find a market with someone making broom handles. Trees like aspen, the lowly popple tree, started to look good to some sawmillers, and they experimented with different uses for this common wood, suffering the derision of other sawmill men.

The Jones Lumber Company sawmill in Wabeno, Wisconsin. Jones was a big logging and lumber outfit headquartered in Appleton, Wisconsin. They logged and sawed lumber all over the U.S. The Wabeno operation came about when the Rusch brothers, one of the first to build a sawmill in Wabeno after the C & NW arrived in 1897, went bust. Jones picked up the pieces and stayed, operating their Phoenix Log Haulers as well as horses. The mill was sold to L. N. Fisher in its later days and operated into the 40's. Fisher rode the train back to Appleton on the weekends. He is noted for being extremely cheap, and often said, "I came to Wabeno to make money, not friends." The Wabeno Grade School now occupies this site.

The Jones Lumber Co., in the foreground, shared this valley with the Menominee Bay Shore Lumber Co. There was friendly competition between the mills and the mill crews, but many people worked at both mills through the years. The farm on the hill is a good indication that this picture was shot in the 1920's, when the homesteaders had made good progress in building their farms.

Below:
This logging camp was built in a stand of hardwoods. Before long, it was in the wide-open spaces. The loggers moved on, leaving the buildings to Mother Nature. It would be at least 70 years before there was another stand of trees on this location that could be harvested for the good of society. On the other hand, the transportation system of the time, the tax laws and the market place dictated that this was the way timber would be harvested. Many of these old camps found further use during the depression years of the 30's. Homeless people moved into abandoned logging and sawmill buildings. Very often, there were numerous items left behind that helped sustain life. An old crosscut could be utilized to cut firewood, and often the stoves were left in the camp as well. Old kitchen utensils and pails were put to use for cooking and gathering berries. Life was hard, but it was life!

The secretary for L. N. Fisher Lumber Co. related to me how Mr. Fisher tried to utilize popple in his mill. Fisher had bought the Jones mill at Wabeno and was trying to extend the length of time he could operate. He bought timber forty miles north at Tipler and hired the C & N W to haul the logs into Wabeno. He also started experimenting with popple lumber, trying to get his customers to buy popple for some of their manufactured goods, like chair seats. He had some success, but other sawmill men thought it ridiculous and some lumber buyers used to address mail to: Popple Lumber Co., c/o L. N. Fisher, Wabeno. By the way, aspen, or popple to use the common name, is a common wood used in chair seats today.

Small trees might not have been appreciated in the early years of the big cut, but were utilized more and more as timber grew scarce. These fellas are cutting small timber, as can be seen by the piles behind them. The oxen are still in use though. Notice the goad with the whip attached to get their attention.

Wabeno, Wisconsin in 1900 was just three years old. It's just a toddler of a town, but think of it, there was naught but wilderness here in 1897! The sounds of saws and hammers must have been a daily occurrence. The Indiana Lumber Company have set up business in the foreground, but the mill was sold soon, and was run by a local interest. The big mills, Jones and Bay Shore are still in the future. The C & N W depot is roughly in the center of the photo, the water tower is to the right, and the log cabin to the right of the water tower was the C & N W land office, where land was sold to lumber companies and homesteaders. Believe it or not, the old land office is in fine shape today and serves as the library.

Looking down the street of Wabeno a few years later, a few more businesses are opened, most of them dispensing liquid refreshments. Quite a few pulls of the tap handle are represented by that wagonload of beer barrels. I interviewed Lawrence Mallow, son of John Mallow, owner of the Eagle Saloon, when he was in his 90's. One of his first memories was crawling under the board sidewalk in front of the saloon, looking for coins dropped by the lumberjacks and sawmill men.

Hotels were one of the first businesses started in the new towns. Sometimes grandiose names like the Milwaukee House were given to hastily constructed hotels that were the family home, hotel, and often saloon. This particular establishment was in North Crandon, Wisconsin, which grew rapidly after the Soo Line railroad was built through in 1887. North Crandon was a switching point between Minneapolis and Sault St. Marie, and train crews spent the night in various hotels.

The Milwaukee House was one of the few saloons to see gunfights. Usually lumberjacks were satisfied to settle their differences with fists or canthook stalks, but a story of a man wounded by a gunshot shows up in early papers, and a story of a shooting was related to me twice, in similar versions, of another shooting in the Milwaukee House. It seems that two gents in from the woods were on a bender, and they had made a few enemies. The sodden duo went into the saloon of the Milwaukee House, located on the right end of the building, and ran head-on into the four men who wanted a piece of their hide. One of the four locked the door and pocketed the key, and, along with the other three advanced on the two victims. Unfortunately for the four belligerents, one of the jacks pulled a pistol and cut loose. In one of the versions, three of the four were killed. In both of the versions, the two drunks didn't bother to get the key to the door, but kicked out the window and went on their way. Nobody was prosecuted, it was a self-defense shooting!

The hardwood era of logging was an awakening for the timber industry, at least east of the Mississippi. The forests weren't forever if they were cut indiscriminately. During the 1920's, some visionary people like R. B. Goodman, who started the community of Goodman around his sawmill in Marinette County, began lobbying the Wisconsin legislature for laws that would make it possible to defer real estate taxes on forestlands until they were cut. The "Forest Crop" laws instituted in the 20's made it possible for Goodman to selectively harvest his remaining timber lands, allowing younger trees to grow into merchantable size trees, and cutting the land again. Goodman started a renewable forest that is still being logged today. The same is true of the Connor Lumber and Land Co. tract at nearby Laona.

> R. B. Goodman tried to build a model community at Goodman, Wisconsin. He owned the houses and all the businesses. One former resident said he even owned the people. Saloons weren't allowed in town, and his people were supposed to find their entertainment at the clubhouse he provided. He was considered a skinflint, was aloof, and didn't have many friends. It is said that when he died, a mere fifteen people attended his funeral. The mill and the town are still operating today, with different owners, of course.

The loggers during the latter part of the big logging era would have been considered pampered by their counterparts who logged in the middle of the 19th century. The "shanty boys" of the pine era lived in hovels compared to the loggers of the 20th century. With the luxury of rail travel, men could even get home to see their families at Christmas. Toward the end of the logging camp days, some of the men brought their automobiles to camp and would go home on the weekends or whoop it up at nearby saloons.

> My father always enjoyed telling about working in a camp near Phelps, Wisconsin when some of the men brought their cars to camp. He was working with two brothers who were huge men, 6'5" tall and all muscle. On a Saturday night, they fired up the old car, owned by his large partners, and went to look for some fun. The larger brother decided they should have a bottle of whiskey to get the sawdust out of their throat. He ordered the car stopped at the very fashionable King's Gateway. Dad and the others in the car explained that this was not their kind of establishment, and should look for a saloon somewhere. "Bullshit!" said the big logger. "There ain't no place too good for a lumberjack." He chose the door to the dining room and entered wearing bib overalls, no shirt, and cork boots. The rest of the loggers crowded into the doorway to see how many men it would take to throw him out. The thirsty jack sauntered through the dining room, wealthy patrons stopped spoons and forks halfway to their mouth and stared, as the floor tiles stuck to the corks of his boots. With a flick of his foot, he freed the loose tiles from his corks. He approached the bar and ordered a quart of Kessler's. The nattily dressed bartender said, " Yes sir!" He left with the bottle leaving a new trail through the floor tile. "See, they were very friendly to lumberjacks," was his reply to my father and the other astonished loggers, as the '27 Chevy roared down the road.

The automobile also made it easier to bring fresh meat into camp, and smoked and salted varieties of the past weren't table fare quite so often. For sure, many was the woods boss who decried the use of the Model T in camp. After all, now the men could get drunk on the weekends, showing up on Monday morning with a hangover, unlike the early days when the jack had to wait until spring for one big bash. This also made for a cash flow problem, as lumberjacks demanded pay on a more regular basis, and not just in the spring.

> Having had the experience of listening to the tales of a good many of these old lumberjacks, I have heard repeatedly of logging companies cheating the men of their pay. Certainly, not all logging companies were crooked, but cheating did exist. Oliver Campbell told of being recently married and trying to get a small farm in operation. Oliver says he worked all winter, drawing only enough pay to barely support his wife and child. He planned to use the money owed him in the spring to buy some young stock, a workhorse and other necessities. In the spring, the owner of the camp left for Alaska, along with the money owed his men. He was never seen or heard from again. Oliver, who was then pushing ninety years of age, said he would still put a bullet in him if he walked through the door.

The old time logging era had begun just before the middle of the 19th century, and by the 1920's, the span of a man's life, it was about over. There were companies that prospered beyond that point, but the big rush was done. No more towns were being created, few mills were built, and many of the little communities that were the source of pride and hope in the future, no longer had enough population to support a store or a post office.

Page & Landeck built a sawmill in Crandon, in 1901, a town that had 800 souls, many of them who lived miles from the town center. In a few years, there were over 2,400 people in the village proper, and in 1908, an entrepreneur named Luigi Parise built the Opera House. It was a building almost a half-block wide, with a notions store on one side of the main entrance and a café on the other side. Apartments were built around the central hall. It was instantly the community center. There were piano recitals, plays by local actors and traveling troupes, magic lantern shows, basketball and graduation ceremonies, and roller-skating. It was the pride of the community.

It is true that towns in the northern timber regions were built quickly, it is also true that big part of them disappeared even quicker. Fire was the bane of all sawdust towns. Fire departments didn't exist, unless you count bucket brigades. The only defense against fire was stalling it long enough to get some of the furnishings and personal belongings out of the building. This is Crandon's Opera House in 1912, four years after it was built, burning to the ground. Notice the wagon piled with items from the building in the foreground. Everyone hoped Luigi would rebuild, but it never happened.

Connor Lumber and Land in Laona, Wisconsin built their company store next to the C & N W tracks, making it easy to unload freight from the rail car directly into the store. The store carried everything imaginable. Harness the horse, dress the kids, feed the family, it was all there. People working for the company had credit at the store, and their needs were met. Workers who overdid the credit, would receive a receipt in their pay envelope instead of a check. It must have been hard for some of the men to drink beer without any cash, and I wonder, did some of the wives use it as a tool to keep their husband sober?

Almost any carelessness could cause a fire in the wooden buildings that made up these sawmill towns. The Connor company store went up in flames in March of 1914. It could have been a chimney fire, a careless smoker, who knows?

What was important to the community of Laona was that the company store was rebuilt within months of good fireproof brick. In this shot, the store is under construction, with the store on the ground floor and the corporate offices on the second floor. The store was equipped with fire-proof safes, and the type of construction told everyone that the Connor Lumber and Land Company was here to stay and offered them a future. The building to the left was also of brick, and was the Hotel Gordon. By the way, the store is no longer a company store, but at the time of this writing, was still selling groceries to the residents of Laona.

Fire destroys the Commercial Hotel in North Crandon. I was told by an early resident that the cook was making apple pies on a summer day when a chimney fire started. The unlined brick chimney split, allowing the fire to escape and burn the building. Remember, the days of gas and electric stoves hadn't arrived, and all the cooking was done with a wood-fired range. As the saying goes, "If you can't stand the heat, get out of the kitchen!"

Wabeno, the town that seemed to sprout like a mushroom, had a business district that was doing fine until the 4th of July, 1914. The town was celebrating, the revelers were out late, but a celebrant who had enough and went to bed early decided to smoke in bed at Ben Slowe's hotel on the east end of the main business block. The result was a fire that spread west and destroyed every building save the stucco-covered drug store on the west end of the block. Eight businesses had no home, including the bank, the opera house, the hardware store, Pichotta's hotel and restaurant, Moore's Buffet, the pool hall and the confectioner's store, and the families that lived above the stores. There was no fire department, and as the paper read, "All the pails in town were brought into use." The people were resilient though. The fire occurred after midnight on Saturday morning. The bank had reopened by Monday afternoon in temporary quarters. Shortly thereafter, Moore's buffet was doing business in the bandstand, Pichottas were selling meals from a tent, and Ben Slowe had set up a "temporary saloon." The next time the buildings were constructed, they were more fireproof, and the more disasters of this type occurred, the closer the citizens of the logging communities came to shouldering the expense of a fire department.

Not all the mills located in the logging communities belonged to big companies. Some of these small mills catered to local needs for lumber and lath, and some brokered lumber through the bigger company. The Todd mill in Crandon was a small mill on the edge of town and remembered by few of the citizens today. Sawmills, big or small, were inherently fire traps, and especially in the days when they were powered by steam, with a constant fire under the boiler and sawdust and wood chips everywhere.

The Todd Mill Fire Crandon Wis

The Page & Landeck Mill in Crandon, Wisconsin, was a modern mill for the day, with the boilers enclosed in a brick building. Even so, the mill caught fire in 1921, and Art Freimuth, a local photographer was on hand to record the event. The mill was owned at this time by Fish Lumber Co.

Just twenty years before, this equipment was hauled in by sledge and horses from Pelican Lake on the C & N W Railroad. The boilers, head rigs, edgers, and the rest of the equipment was drug for twenty miles or so over rocks and swamps so the mill would be operational when the C & N W spur was built in 1901. This mill was never rebuilt, but an employee of the company, Forrest Himes, bought the wreckage in 1923 and constructed a mill just south of this sight. The mill operated for about 40 years, and it too burned twice during this period, but Himes rebuilt. The mill was smaller, and the logs were brought in by trucks.

Steam power was considered a modern miracle in the 19th century and well in to the 20th century. Steam power was not without its hazards though, and many a boiler blew up. On March 17, 1908, the Kempf planing mill blew up in Crandon. Three men were killed, several were injured. It was no doubt the result of the operator not watching the water level in the boiler.

Below:
The boiler from the Kempf planing mill, laying on the fence with the flue pipes exposed, went airborne for two blocks, bouncing off the roof of the house before coming to rest.

As the lands were cut off, those in the know who could see the big picture, the big holders of these cutover lands, tried to divest themselves of their holdings. Many before the calendar turned over to 1900, and almost all lumber companies after the turn of the century formed land companies.

The virtues of homesteading your own piece of ground were touted to people working in mind-numbing factory jobs. Fresh air in the wilds of Wisconsin or Michigan, understandably, had a great appeal to someone breathing coal smoke in Pittsburgh or working in a brewery in Milwaukee. The lumber/land companies promoted the freedom and good health that was available to enterprising men and their families in what was to become the finest dairy land in the entire nation.

Brochures were distributed in cities and existing farm areas that extolled the virtues of the logged-off land, and inexpensive train fares brought the hopeful north to the next big farming center. One of the favorite tricks was to show the land to prospective buyers in the winter. The rolling hills covered with snow and small trees looked like the perfect place to raise a family, and the pioneer spirit bubbled over. The land companies had easy payment plans, and many bought on the spot before returning to their labors in the factories with a light heart. In the spring, the family would arrive with their possessions in a rail car and their pockets full of hope.

Homesteaders lived a rough life, but many would tell you that it was better than the old country or the steel mill in Pittsburgh. Most gave up on farming, but not all. Many nice farms were created by people like this, who came from eastern Europe and literally carved out a new life in the northern woods.

In the days before game laws and, for that matter, in the days after game laws, nature added to the table of the homesteader. Why butcher the cows when deer ran wild through the slashings? This fellow, a homesteader near Nelma, Wisconsin, was a successful hunter. If you look closely, you will see two fawns and two does hanging besides the two nice bucks. He is well equipped with a modern Winchester rifle, and was probably the envied by many of his neighbors.

A stump puller was a luxury not enjoyed by many of the homesteaders who cleared the land after the loggers left. Most stumps were removed with grub hoes, shovels, axes and dynamite. Horses were used to pull the fragments from the ground. Dynamite was a great aid to stump removal, but caused a lot of injuries to settlers untrained in its use. The University of Wisconsin ran a special train called the Land Clearing Special to the cutover teaching homesteaders the safe way to handle dynamite and remove stumps.

The social welfare net in the 20's was the county poor farm. Elderly people with no families were sent to the poor farm, and families down on their luck. Residents of the poor farm raised their own food and supplied whatever skills and work they were capable of in return for their keep. These men are clearing land for the poor farm in Forest County, Wisconsin.

The reality was rolling hills covered with stumps and rocks left by the last glacier, and hidden by the snow. Surely, some of these families must have got on the next train and went back where they came from, but many, probably most, attacked the rocks and stumps with a vigor that can hardly be imagined today. Some of these farms still exist today, and most of the bigger successful farms are an amalgamation of several of the first farms. A drive through the countryside will tell the story. A grove of lilacs and apple trees in the middle of a hay field is the mute memorial to a family that cleared the stumps, picked the rocks and tried to make a comfortable home and prosperous business. Their fields are now part of a bigger farm that somehow managed to survive the loss of the lumber company sawmill and camps that bought much of their produce and the Great Depression. Of course, railroads made the production of crops like potatoes marketable to the population centers, and in some areas, farms did prosper. In other areas, the railroad was too far away, and the logging camps were the only source of income. When the timber was completely gone, the small, isolated homesteads didn't have any viable markets. The fact was that all of the backbreaking work had been in vain, and families moved away, sometimes to the very factories they had left years before. The great forest that had been so recklessly cut started to grow back on the edges of the cleared fields, and in some instances has completely taken back the homestead, save for some old apple trees that have survived marauding bears with a sweet tooth, and clung to life through the decades.

The homesteader raised much of their food, but there were items that were impractical to raise. Flour and other common staples, clothing, nails, tools, and a host of other items had to be purchased. They also had to come to town to sell produce. This lady put on her good bonnet to come into Wabeno back around 1900, and the slow but dependable oxen got her there.

> *When I was still in the logging business, I was logging an eighty-acre tract of marginal timber, when an old timer showed up and started to reminisce. He told me that as a young child he had visited this site with his father. A farmer, he said, was trying to raise sheep on this rocky hillside, and he had enclosed the entire eighty with a woven wire fence. I doubted that a forest, albeit a poor quality one, had been able to grow that much in those few years. That afternoon, as I picked up a bundle of birch pulpwood with the loader of my skidder, I hooked a woven wire fence that was lying under the duff of the forest floor. When I raised the pulpwood, the wire popped up from the forest floor for a hundred feet each direction, throwing sticks and fragments of old fence posts into the air, a testament to the memory of my visitor and the resilience of the forest.*

Personal transportation was a problem in the early twentieth century in the cutover. Most people couldn't afford a car, and if they could, roads weren't plowed in the winter and were muddy sinkholes much of the rest of the year. The railroads didn't like it, but some of the homesteaders living in outlying areas used the rails for personal use. Speedos like these moved right along winter or summer. Bicycles were also equipped with an arm and a third wheel to ride the rails to town.

What happened to the big sawmills of the past? Some were torn down and moved west, where timber for the taking was still available, and many were sold to the insurance company. Fire was the enemy of the sawmill owners while the timber was still standing, but became a handy friend when there were no more logs to ride the carriage. Big sawmill fires were common, and they often signaled the end to a prosperous community. The owners usually moved on with an insurance check in their pocket.

> When the Boom Lake Lumber Company stopped operations in Rhinelander, Wisconsin in the early 40's, they were proud of the fact that they didn't sell their mill to the insurance company and said so.

There was human wreckage as well. Middle aged to old men, who knew no other trade than logging, found themselves without a logging camp to live in, and many had a problem with the whiskey bottle. Some of those old jacks who were still capable of working in the woods applied their talents to cutting pulpwood for the growing paper industry. The big logging companies had no interest in spruce swamps, but the paper mills did. In the decades that had passed since the first logging ventures, the older cutovers had grown back with aspen, and a market developed for peeled popple at the pulp mills.

> Popple was peeled in the spring of the year when the sap comes up. The bark can be easily slipped from the trunk of the tree using a spud. A spud could be made from an old car spring. The popple peeled from early May until around the 4th of July, when the bark tightened up. Then the peeled trees were cut to length and bunched in piles along brushed out trails. After the bunching operation was done, horses pulling drays were loaded with the peeled popple and the wood was skidded to roadside for hauling with trucks to a railhead. The hauling could last all winter. After a short lay-off in the spring, the cycle would start over.

Some of the old jacks ended their days helping out on a farm, a job they wouldn't have done at gunpoint during the big logging era. Some just wandered off and were never heard from again. The logging and sawmill industry wasn't noted for its pension plans. A good man in the woods was appreciated, but after his robust days were over, nobody worried about his well being, including many of the old lumberjacks who didn't seem to worry much about their own future. Indeed, most of the small town cemeteries have large blank areas of mowed grass that is the final resting place of many of these men who didn't have families, any money to cover their funeral or to buy themselves a tombstone. Their remains will forever reside in these potter's fields in anonymity.

> Growing up in a small town in northern Wisconsin, I was able to see first-hand the old lumberjacks who were reaching life's end in the 1950's. Many were drunks who lived off a tiny social security check. Some did live in a cabin on farms, where they worked for their room and board. There were also "shackers." Shackers were old lumberjacks who cut pulpwood for one logger or another and lived alone in a shack of their own construction on the job site. Winter evenings had to be extremely long, when the rest of the crew left at 4:00 o'clock, and the shacker, usually in an 8' by 12' tarpaper shack, was the only soul on the job, miles from town. And maybe some of them preferred things that way, being a lone wolf type by nature, but I suspect most lived a sad life.
>
> One of the stories I grew up with concerned a half-breed Indian named Oscar. As a kid, Oscar scared the hell out of me. He was big and rawboned, and he was bald as a bowling ball. He had fierce looking eyes and didn't seem to smile at all. He was living in a shack in the woods, cutting spruce for Maggie Sjoquist. Maggie, who's real name was Magnus, was a Swede who spoke with a strong accent and was well liked by everybody in our small town. Oscar was known to get to town every once in awhile and get stone drunk, which he did one weekend in the early 50's. On Sunday evening, it started to snow, dropping a foot or so of the white stuff. On Monday morning, Maggie went out to his job site and fired up his dozer to plow the road out for the rest of the workers who would be coming to cut pulpwood later in the morning. (Loggers by this time didn't live in camps, and drove from home to work every day.) As Maggie was rolling the snow out of the road with the dozer blade, he saw a foot sticking out of the snow bank. He proceeded to dig out old Oscar, who had laid down in the snow when he was walking back to his shack after filling himself with "antifreeze." He took Oscar to his shack, built a fire and thawed him out. Oscar didn't seem the worse for wear, and, according to Maggie, was cutting pulpwood with his one-man saw that afternoon!

The Lake States sawmills and logging camps had roared with enthusiasm decade after decade. Cities were built, fortunes were made, but the inexhaustible supply of timber had run out. Certainly, there was still a timber industry, and lumberjacks still lived in the woods and cut trees, but the business had changed and was a shadow of itself by the 1920's. The scattered remnants of the big woods would be cut in the years to come, but more of that timber, every year, would be brought to a railhead or sawmill on the back of a truck. The sawmills, too, would be smaller in size and had to be more efficient. The big mills that would spawn a community were a thing of the past. The internal combustion engine as a source of power would replace steam power, and eventually, diesel engines were replaced by electric motors.

Like any vibrant industry, men of vision would come forward with new inventions that would make jobs less labor intensive and more efficient. Over in Eau Claire, James McDonough would build saws and other sawmill machinery that was more dependable. Working as a millwright at one of the huge pine mills on the Chippewa River, McDonough and his inventions began to demand more and more space in the mill. By the time the pine ran out and the sawmill had closed, McDonough and his, by now, famous machinery was using the lion's share of the space in the mill.

Fifty years later, at the family sawmill in Newald, Wisconsin, Francis Cleereman would return from World War II and invent the first successful automated sawmill carriage. Both McDonough and Cleereman Industries are still in business and supplying quality machinery to the sawmill industry worldwide.

Did the Lake States suddenly stop supplying lumber and paper to the world? No, but the numbers were smaller, and the products were different. The big tracts of virgin timber were gone, and the remains were a source of national shame, as well it should be. While modern day environmentalists like to blame the logging industry, including, undeservedly, today's logging industry, for the huge waste and large theft of the public's resources, the government and the politicians should also take blame, and maybe the majority of the blame. The public treasure chest could not have been plundered without the complacency and, often, the complicity of the people holding political office. Very few had any vision for the future and the needs of future citizens for healthy, productive forests. Had the timber held out a few decades longer, the Great Depression would surely have put the brakes on the big cut.

The Great Depression did come, however, and the brakes were put on almost everything. The fledgling U.S. Forest Service, charged with creating and expanding the national forest system, found tax delinquent lands available and local governments, usually with empty coffers, ready and willing to cooperate in establishing a new national forest. The Civilian Conservation Corps not only gave young men a job and a purpose in life, it also gave national forests all over the nation an inexpensive labor force to replant trees, build fire lanes and ultimately protect the invaluable watersheds that had been damaged in the cut and run logging days. Tax incentives for people to grow timber and pay the taxes after harvesting the logs years later, made it economically feasible for sustainable logging. The forests are back, but that is another story, for another time.

Chapter 3
Padus, Profile of a Sawdust Town

The Great Lakes region is dotted with the remains of small communities that took root because an entrepreneur had a stand of timber and a sawmill. It must have been exhilarating to found a new community in the wilderness. Certainly, thousands of people participated in the endeavor throughout the Lake States.

When the Chicago & Northwestern Railroad pushed a line north from Green Bay to eventually connect with Iron River, Michigan, the opportunity to go north and make a new life was irresistible to a great many people. John E. Hammes, a sawmill man from Green Bay was lured north. He bought a tract of timber between Laona and Wabeno, Wisconsin. He built a sawmill, homes for workers, a boarding house for the single workers, and a big house that served as his home, company store, post office and a hotel for travelers on the railroad that ran a few feet from his front door. In addition, the Holt Lumber Co. established a heading mill at Padus. (A heading mill made the ends for barrels, which were used to ship everything from crackers to nails.) Presumably, Hammes Lumber Co. sold some of their lumber to Holt at the heading mill, a handy market in the back door of his sawmill.

The heart of the community was the general store, post office and hotel, which also was the home of John Hammes and family. The boarding house is directly behind the store, and the company houses are lined up beyond the boarding house. The C & N W stopped at the front door.

If you could turn back the clock and shop in the Hammes store…..John and his wife Anna would greet you and be able to fill many, if not all of your basic needs.

The post office was on the other end of the counter in the Hammes store.

The Hammes mill provided most of the employment for the community, in fact, it was the reason there was a community. Hammes also planed his lumber, and made moldings, spindles and other after-market products.

The Holt Lumber Co. owned and operated this heading mill at Padus. Headings were the ends for barrels, the container that was in common use at the time. Cardboard put a lot of cooperages out of business as the paper industry gained prominence in the Lake States. But in 1910, the barrel was used for shipping almost everything in bulk.

A number of years ago, I wrote and designed a brochure for a tourist railroad that was trying to make a living on the abandoned C & N W rail line that passed through Padus. I found a few old photos of the Hammes home and a few shots of the town when it was still a town. The brochure generated interest from the people who volunteer their time to run and maintain the Wabeno Logging Museum, and one of them told me they had glass slides that they thought came from Padus. The glass slides were actually glass negatives taken by an unknown photographer with an old 4x5 view camera. Joe Irocky, a retired photographer from Chicago, had brought his 4x5 enlarger with him when he moved to the Northwoods, and we soon had several sets of 60 8x10 photos of Padus, and a wonderful look at what life was like in the old ghost town.

John Hammes with a young worker in the mill yard of the Hammes mill.

Hammes was decking logs in his yard without the benefit of a jammer. The decking lines are strung over the pile.

John Hammes didn't have a huge sawmill, but he did make moldings and house trim, giving an extra value added to much of his lumber. He also operated a 6000 tree sugar bush on some of his property, making maple syrup and maple sugar one of the products manufactured in his town. His store and post office also contributed to the family income, as did the guests who rented a room in his hotel. Hammes didn't become wealthy as the mainstay in his community, though he did live comfortably by the standards of the day. The combination house, store and hotel was heated by a furnace in the basement, a luxury in the days when most central heating was found only in public buildings. He could also afford a better-than-average automobile and was able to clothe and feed his family quite well.

Falling a hemlock for the Hammes mill. Hemlock was bypassed by the earliest loggers, except for building temporary barns, sheds, etc., but after the pine had been logged off much of the land, species like hemlock, tamarack, and spruce became more important as building materials. In fact, I would bet that most of the old homes in the cutover region are built of hemlock. While it splits easily, and the butt log usually had wind shake and was left behind, hemlock makes durable, strong lumber.

Helen Hammes Chinnock, granddaughter of John Hammes, remembers a big touring car bought by her grandfather after the roads were developed to Padus. She says that John brought her a piano in the backseat of the car after a trip to Green Bay, and on another trip, he brought home a Shetland pony in his backseat.

John Hammes had a son, named John Hammes, who married, had two daughters, and lost his wife when they were very young. He brought his two children to live with his father and sister Annie. Helen, the youngest daughter, has lived into her nineties, and is still spending summers on nearby Trump Lake. Helen Hammes Chinnock had a very successful career as an educator, and recalls the years she spent with Aunt Annie with fondness. She has been a great source of information about what life was like in Padus. In this photo, Helen is delighted with a litter of newborn puppies.

Helen Hammes Chinnock in June of 2001, at her cottage on Trump Lake, just down the road from the town her grandfather founded. Helen was 92 when this photo was taken. She was born in Padus, in 1909, in her father's house.

Padus was not an isolated island of civilization in the wilderness. The C & N W stopped in his front door, and the City of Green Bay was a short 90-mile train ride away. The railroad connected him to his markets, his extended family in Green Bay and whatever cultural pursuits he might miss. Families filled the company houses and some families built homes of their own, a few of which are still occupied. Wabeno was a few miles away and easily accessible by horse and carriage, horse and sleigh in the winter, and automobile, in later years.

Before roads were built and after they were built, but not plowed in the winter, the bobsled was a common mode of transportation. The picture of this family outing in a bobsled was taken in the row of company houses that Hammes built for his employees.

Not all residents of Padus lived in company houses. This old gent has opted for the rustic model. Sure, it looks like rough living, but back then, he may have been quite satisfied to have his own cabin, a rifle, and from the looks of the muddy ground, he has made it to spring and hasn't run out of firewood.

Was Padus special or different than other sawdust towns in the upper Great Lakes region? The answer is no, but the fact that it was typical and that a comprehensive photographic record exists makes it somewhat special.

The local legend says that Padus was named for a dog that followed a traveling preacher into the fledgling town. When the preacher left, his dog decided to stay, no matter how much the preacher coaxed the animal. Maybe Padus found a person in town that he liked better than the preacher or maybe he just couldn't stand to listen to one more of the man's sermons, either way, the town was named in honor of the dog that adopted it, and it retains the name today.

Padus would have been formed after the C & N W laid their rail, which was 1900. By the 1920's, the pickings must have been getting slim, because in 1932, there was no longer enough business to run the post office, and the federal government closed it down. Even if there had been enough timber to run the mill at full tilt, it is likely that the Great Depression would have closed its doors. Eventually, people moved on, John and his wife Anna passed away and were buried back home in Green Bay. The boarding house and the company houses stood for several decades, but without repairs and upkeep, they eventually collapsed and rotted back into the soil. In time, the only Hammes left in Padus was Annie, John and Anna Hammes' daughter.

Annie spent most of her life on the frontier, mixing with lumberjacks and saw millers who lived hard and talked rough. She acquired the same language skills as the people with whom she spent most of her time. Annie never married. According to a niece, Helen, that she helped raise at Padus, she had a suitor from Green Bay who intended to marry the rather attractive saw miller's daughter. After riding the train up from Green Bay on a surprise visit, he walked into the store and caught Annie turning the air blue with her colorful language. He turned on his heel, got back on the train and never returned. It was probably his loss. Annie did an admirable job of raising two nieces, caring for her aging father, and was a hard-working, self-sufficient woman, who valued education and was well read.

Most of the people who still remember Annie never knew her when she was a young woman. They remember the solitary figure that lived in the big house out at Padus, who kept a number of dogs, and had the dubious honor of being the last resident of the original town. In truth, Annie's mind didn't hold up well, living alone with the ghosts of the old place. She was put in a nursing home in the mid-sixties and died a short time later. The times had passed her by, or she just didn't want to participate in a modern world. She was noted for her sense of humor and her cussing skills that rivaled the best efforts of the most colorful teamsters.

Annie Hammes as she appeared about 1910.

Previous page and above:
Very often, men, and sometimes women, posed with their favorite firearm. It would be strange today to grab your favorite gun when your photo was to be taken, but in the early days of the last century, firearms were not only a means to get fresh meat, but was one of the primary forms of entertainment. Guns were treasured belongings.

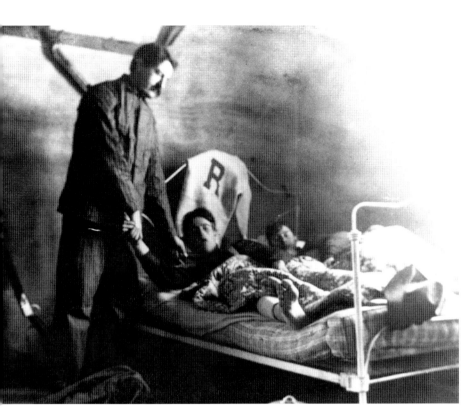

The railroad connecting Padus to the lumber markets also connected people from the outside to Padus and the northern deer herd. With a hotel to stay in, hunting deer in the northern woods became less of an ordeal. The hunters in these photos appear to be guests in the Hammes Hotel. Maybe they have spent a little too much time the evening before at the saloon and dance hall just down the road from the hotel. But, it's daylight in the swamp, and time to hunt! It should be safe to assume that the photographer wasn't shot seconds after this photo was taken, as the photographic record continues.

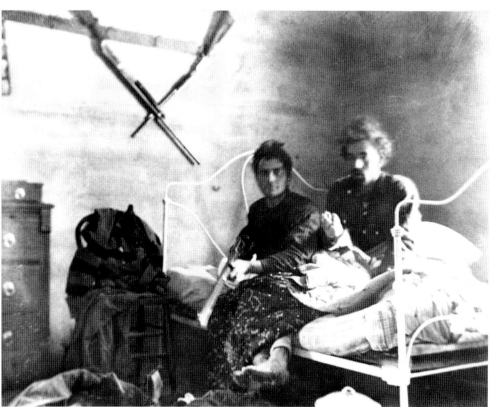

Dogs were often used to hunt deer in the old days. It was eventually made illegal, but the practice continued for years in spite of the law.

Frozen deer can stand again. Annie and Helen have sneaked in to the picture. The C & N W tracks are in the background.

Hunters and their hound have taken a small bear, and the photo is an excuse to clown around.

Young John has just caught a rather large northern pike through the ice, big enough to warrant a photographic record of the catch.

Padus was built close to three lakes, and lakes figured prominently in the recreational time of the residents. These lovely ladies are showing quite a bit of ankle for the day.

Young people, like young people of today, were drawn to the lakeshore for picnics, courting, and other things young people have always done, including goofing off.

Be careful when walking on a log! The old river drivers knew that, but these two ladies apparently didn't. The girl on the left was the schoolteacher in the Padus one-room school, and it's a good thing her students weren't there to see her take the plunge. By the way, the photographer, whoever he or she was, should get some recognition for getting these shots with a cumbersome view camera and glass plates.

Padus is hardly a ghost town today. A number of people maintain nice homes there, but it is no longer a village, and is part of nearby Wabeno. The post office was decommissioned in 1932, when Annie lost her job as postmistress. The mill shut down in the late 20's. One by one, families left Padus looking for more opportunity. The Hammes home, store and hotel lasted into the 1990's, unlived in and showing the effects of vandalism and age, finally succumbing to fire, either from arson or lightning, nobody knows. While the forest claims back the land, these photos remain as a reminder of a happy time, when people came together to build a community in the woods, sharing the good times and the bad, like many others of the era.

It was time to have a smoke with a Native American friend. Annie is in the background with an expression that seems to say, "There's work to be done!"

A young family, possibly in their first home, might be good guess for this shot. The boards around the bottom of the house are holding in sawdust that insulated against the cold winds of winter. The bucksaw to the left of the window would have kept the stove supplied with wood, and would truly heat the woodcutter twice.

Daily living took its toll on women, especially with two young children. The fatigue is clearly visible in this mother's eyes.

I wonder how much the women of Padus and women everywhere silently cursed the long skirts style dictated they wear. Keeping hemlines clean in a world covered with mud had to be a never-ending job. Is it any wonder that they embraced the short skirts of the Roaring 20's!

It must be Sunday, or these people would be working, instead of enjoying a summer day by the company houses. The young boy is playing soldier, and unbeknownst to him, he may have to do it for real when 1917 rolls around.

A family living in Padus. When you get right down to it, that's why places like Padus were built-to support families. If the going was rough at times, well, it made you a tougher individual. Former residents of this little town in the timber always talk with fondness of their years at Padus. Maybe it toughened them to the trials that life can bring, and maybe they are a little proud of the fact that they helped create a town in the big woods.

A solitary figure stands in the snow behind the Hammes house, and if it is Annie Hammes, it may be a harbinger of her future.